Y0-BYC-629

Matrix Algebra As A Tool

Ali S. Hadi

Cornell University

7379

An Alexander Kugushev Book

Duxbury Press
An Imprint of Wadsworth Publishing Company

I(T)P™ An International Thomson Publishing Company

Belmont • Albany • Bonn • Boston • Cincinnati • Detroit • London • Madrid • Melbourne
Mexico City • New York • Paris • San Francisco • Singapore • Tokyo • Toronto • Washington

Editorial Assistant: Janis C. Brown
Production: Gary Mcdonald
Marketing Manager: Joanne M. Terhaar
Print Buyer: Karen Hunt
Permissions Editor: Peggy Meehan
Copy Editor: Paul Green
Cover: Janet Bollow
Printer: Malloy Lithographing

COPYRIGHT © 1996 by Wadsworth Publishing Company
A Division of International Thomson Publishing Inc.

I(T)P The ITP logo is a trademark under license.
Duxbury Press and the leaf logo are trademarks used under license.

Printed in the United States of America
1 2 3 4 5 6 7 8 9 10—02 01 00 99 98 97 96

This book is printed on acid-free recycled paper.

For more information, contact Duxbury Press at Wadsworth Publishing Company:

Wadsworth Publishing Company
10 Davis Drive
Belmont, California 94002, USA

International Thomson Publishing Europe
Berkshire House 168-173
High Holborn
London, WC1V 7AA, England

Thomas Nelson Australia
102 Dodds Street
South Melbourne 3205
Victoria, Australia

Nelson Canada
1120 Birchmount Road
Scarborough, Ontario
Canada M1K 5G4

International Thomson Editores
Campos Eliseos 385, Piso 7
Col. Polanco
11560 México D.F. México

International Thomson Publishing GmbH
Königswinterer Strasse 418
53227 Bonn, Germany

International Thomson Publishing Asia
221 Henderson Road
#05-10 Henderson Building
Singapore 0315

International Thomson Publishing Japan
Hirakawacho Kyowa Building, 3F
2-2-1 Hirakawacho
Chiyoda-ku, Tokyo 102, Japan

All rights reserved. No part of this work covered by the copyright hereon may be reproduced or used in any form or by any means—graphic, electronic, or mechanical, including photocopying, recording, taping, or information storage and retrieval systems—without the written permission of the publisher.

ISBN 0-534-23712-6

Dedicated to

Mr. Fawzi Haleem, my primary school teacher, whom I still vividly remember. He has inspired me and provided me with a role model for becoming a teacher;

and to

All the people (my extended family) of Saft El-Noor,[1] the still tiny village in upper Egypt in which I was born and raised.

---

1 Saft El-Noor is translated as *source of the light*. It was without electricity, but the name certainly suffices.

Preface

I teach intermediate and advanced *applied* statistics courses, such as applied regression and multivariate analysis, at both the undergraduate and graduate levels. Although most of my students are not statistics majors, their various fields of study demand many mathematical tools. Matrix algebra in particular is necessary for a thorough understanding of research methods courses in other disciplines.

Although the students in my classes are highly motivated, they often lack the mathematical skills needed to understand the material. To specify matrix algebra as a prerequisite would be an easy solution for me as a professor but an unsatisfactory one as an educator, because it would prevent or discourage many students who should, on the contrary, be encouraged to take such courses. Therefore, rather than specify matrix algebra as a prerequisite, I have tried two alternative solutions. The first was to present a brief review of the necessary matrix algebra tools and refer students to one of the many standard books on the topic. Matrix algebra books, however, are typically reference books written at a higher mathematical level, and are inaccessible for nonmajors. As a result, many students dropped out, and many who stayed were very confused.

Alternatively, I tried devoting a good portion of the course to matrix algebra. The obvious disadvantage of this solution, however, is that it does not leave enough time to cover the statistics topics that are the real point of the course. I finally compromised, devoting a relatively short time to matrix algebra and then referring students to some lecture notes I had specifically written for them. These notes – expanded, revised, and reorganized – evolved into this book.

The main purpose of the book is to present matrix algebra as a tool for students and researchers in various fields who find it necessary for their own work. It is intended to provide a basic understanding in language they can easily understand. The book's informal style has forced me on occasion to be mathematically imprecise, but this seems to me a small price to pay for increased accessibility. Some students will be able to read this book on their own, and much of it will require little instruction; the remaining part, however, may require a moderate amount of instruction. I hope that even students with prior courses in matrix algebra will find this book helpful.

The book can be used as a supplement to many courses, such as applied regression analysis, applied multivariate analysis, econometrics, biometrics, and other research methods courses in various disciplines such as sociology, psychology, economics, accounting, finance, marketing, and so on. The book may also serve as the primary text for a short course in matrix algebra that might be offered concurrently with the above-mentioned courses, for instance, or during the summer semester or other open periods in the academic calendar. The book may also serve as a text in a more rigorous matrix algebra course, although this was not the original intention in writing the book. In this case, the instructor may wish to provide proofs of some results and assign the proofs of other results as exercises.

I assume no previous acquaintance with the subject. Except for a very brief mention of the word *derivative* in Section 8.3 (which is an application section), calculus is not invoked. Although the book addresses many advanced topics (e.g., Section 6.3 and Chapters 8–11), these topics are covered at an elementary level and are illustrated by numerical examples.

As the title indicates, the emphasis is on practical rather than theoretical considerations, and much of the book is devoted to applications. Indeed, Chapters 8, 10, and 11 are devoted entirely to applications. Matrix notation, vocabulary, and concepts are emphasized. Although some of the results are derived, nearly all are stated without proof. Rather, the emphasis is on the interpretations and implications of the results and, in lieu of proofs, the results are illustrated and verified by examples.

Although there is a great temptation to do so in matrix algebra, I have tried not to emphasize matrix computations. Numerical examples unavoidably involve some calculations, but the examples are primarily designed to illustrate the concepts rather than matrix manipulations. For this reason, methods which are purely computational are relegated to an appendix to the chapter in which they appear. I use elementary methods for calculations, which are simple and easily understood but are not necessarily preferred in terms of computational efficiency or numerical accuracy. In most cases, computations are illustrated using small matrices. Most real applications, however, involve larger matrices, many of which are not practical to compute by hand. Whenever possible, professionally written computer programs should be used for matrix calculations in real-life applications. There are several main frame and personal computer packages aimed specifically at matrix calculations. Even spreadsheet programs now include some basic matrix calculations. I do not refer to any specific software or hardware because a general book like this should not be tied to specific software or hardware. Therefore, instructors may use any software package of their choice.

Geometry and matrix algebra are intimately connected, and matrix concepts can often be easily understood by relating them to their geometric counterparts. I tried to make this connection as early as possible in the book, beginning with Chapter 4.

Exercises are given at the end of each chapter. The exercises are not really optional, as many of them involve concepts supplementary to, but not covered in, the text. Readers should be able to do the computations by hand for all small matrices; otherwise all numerical work should be done by a computer. Also, readers are urged to substitute their own real data for the small matrices in the exercises and perform the calculations using a computer.

Vocabulary and concepts are defined and explained when they are first encountered, and key words are italicized. Some material, marked by an asterisk, is either more advanced, more theoretical, or more difficult to read. These parts can be skimmed or skipped at first reading. Except for Section 8.2 and the parts marked with asterisks, the first nine chapters constitute a minimum background and should not be skipped. The remaining material consists of either applications or advanced topics that can be read when the need arises. Nevertheless, it is recommended that the chapters be read in the order they are presented.

Two other chapters I originally intended to include, on other matrix decompositions (e.g., Cholesky, Q-R, and L-D-U) and on basic matrix differential calculus, were omitted as beyond the nature and scope of this book.

I am grateful to many of my students for the comments they have given me and to William Gould (Stata Corporation) for sharing with me his own handwritten notes on some topics of matrix algebra. Paul Green, Gary McDonald, and Mary Roybal have done a superb job editing the manuscript. I thank the following colleagues at Cornell University for reading and commenting on an early draft of the book: Jody Enk, Robert Hutchens, Matthew Hutcheson, Hyeseon Joo, Shayle Searle, Paul Velleman, and John Walker. I have also benefited from the comments of many colleagues at other universities. I am indebted to Gulhan Alpargu (McGill University), Bruce Barrett (University of Alabama), Samprit Chatterjee (New York University), Daniel Coleman (Carlos III University of Madrid), Mohammed El-Saidi (Ferris State University), Jose Garrido (University of Concordia), J. Brian Gray (University of Alabama), Brian Jersky (Sonoma State University), Robert Ling (Clemson University), Alberto Luceño (Cantabria University, Spain), Hans Nyquist (University of Umeå, Sweden), Daniel Peña (Carlos III University of Madrid), Jaime Puig-Pey (Cantabria University, Spain), Karen Shane (New York University), Simo Puntanen (University of Tampere, Finland), Gad Saad (University of Concordia), Gary Simon (New York University), Jeffrey Simonoff (New York University), Mun Son (University of Vermont), Lenn Stefanski (North Carolina State University), George Styan (McGill University), and Mohammed Youssef (Norfolk State University). Parts of the book were written while I was visiting the Department of Applied Mathematics and Computing Sciences of the University of Cantabria. I am grateful to Enrique Castillo, who made such a visit both enjoyable and productive.

Ali S. Hadi
Ithaca, New York
June 1995

Contents

1 Introduction

1.1. Why Matrix Algebra?

Nowadays, more and more applied scientists use statistics in their work. For this reason, more statistics courses are now required of students in almost every field of study. Matrix algebra is an important tool for a thorough understanding of many statistical techniques, such as multivariate analysis and regression. In fact, you will soon discover that matrix notation is not only necessary for us to fully understand many statistical techniques, but it also greatly simplifies the formulas and concepts of those techniques.

I believe that good and thorough courses on multivariate analysis and regression analysis cannot and should not be taught without matrix notation. We may be able, for example, to write formulas and understand concepts in the simple regression case without matrix notation, but it is a hopeless task to do the same in the multiple regression case.

Matrix notation is also useful in many other fields. Whether you are an economist, a social scientist, a biologist, a chemist, a physicist, or an engineer, knowledge of matrix notation will help you read and better understand journal articles in your own field. In summary, regardless of your own area of study, I believe you will benefit by reading this book.

Coverage. Clearly, one has to be selective when writing a book of this level, size, and type. I believe this book contains (nearly if not all) the material you need for taking intermediate and advanced courses such as applied regression analysis, techniques of multivariate analysis, econometrics, biometrics, and other "research methods" courses in various disciplines such as sociology, psychology, economics, accounting, finance, and marketing. The book also contains some advanced topics (e.g., Section 6.3 and Chapters 9–11). These topics are covered at an elementary level, however, and are illustrated by several numerical examples.

The emphasis in this book is on matrix notation, vocabulary, concepts, and manipulations, rather than on matrix theory or calculations. Proof of matrix theory is usually of interest to mathematicians, numerical analysts, computer scientists, and statisticians. If your interest is in the theoretical foundation for matrix algebra or in its calculations, I suggest reading one or more of the following books: Stewart (1973), Noble and Daniel (1977), Searle (1982), Graybill (1983), Strang (1988), and Golub and Van Loan (1989). Styan and Puntanen (1994) give a very comprehensive and impressive list of books on matrices, inequalities, and related topics.

Matrix concepts and operations are illustrated by simple examples. Calculations are illustrated using small matrices because calculations for larger matrices are not feasible to do by hand; they are usually and more conveniently done by computers.

As mentioned in the Preface, it is recommended that the chapters be read in order. Some parts, marked by asterisks, are either more advanced, more theoretical, or more difficult and can be skimmed or skipped at first reading. The first nine chapters (with the exception of Section 8.2 and the parts marked with asterisks) constitute a minimum background and should be read by all readers. The remaining material, consisting of either applications or advanced topics, can be read when the need arises.

Let us now begin by introducing some definitions and notation, then conclude this chapter with a discussion of some special types of matrices.

1.2. Some Definitions and Notation

1.2.1. What Is a Matrix?

A *matrix* is an ordered rectangular array of elements. These elements could be numbers or symbols (like x and y) that represent numbers. The numbers can be either real or complex.[1] In this book we are concerned only with real matrices. However, most of the results for real matrices still hold for complex matrices. For example, suppose we have three factories at which certain goods are produced. These goods are then transported from each factory to each of four cities. Table 1.1 shows the shipping costs (in dollars per unit) from each factory to each city.

Table 1.1. Per Unit Shipping Costs From Each Factory to Each City

	City 1	City 2	City 3	City 4
Factory 1	5	2	1	4
Factory 2	2	3	6	5
Factory 3	7	6	3	2

These shipping costs can alternatively be written in matrix form as

$$\begin{array}{cc} & \begin{array}{cccc} \text{Col. 1} & \text{Col. 2} & \text{Col. 3} & \text{Col. 4} \\ \downarrow & \downarrow & \downarrow & \downarrow \end{array} \\ \mathbf{A} = & \left(\begin{array}{cccc} 5 & 2 & 1 & 4 \\ 2 & 3 & 6 & 5 \\ 7 & 6 & 3 & 2 \end{array} \right) \begin{array}{l} \leftarrow \text{Row 1} \\ \leftarrow \text{Row 2} \\ \leftarrow \text{Row 3} \end{array} \end{array} \tag{1.1}$$

This matrix, which we call **A**, say, has three rows, representing factories, and four columns, representing cities. Note here that the order in which the numbers are arranged in a matrix is important because the first row represents Factory 1, the

1 A complex number is a number which can be represented by $a + b\sqrt{-1}$, where a and b are real numbers and $b \neq 0$.

second represents Factory 2, and the third represents Factory 3. Similarly with the columns, the first column represents City 1, the second represents City 2, and so on. The number of rows and the number of columns of a matrix, in that order, are called the *order* of the matrix. Thus, the matrix **A** above is of order 3×4 (read as *3 by 4*).

As another example, suppose we take a sample of five children and, for each child, we measure three characteristics: age in months, weight in pounds, and height in inches. The 15 measurements are given in Table 1.2. These measurements can be extracted and arranged in a 5×3 matrix:

$$\begin{array}{ccccc} & \text{Var. 1} & \text{Var. 2} & \text{Var. 3} & \\ & \downarrow & \downarrow & \downarrow & \\ \mathbf{X} = & \left(\begin{array}{c} 1 \\ 2 \\ 2 \\ 3 \\ 4 \end{array}\right. & \begin{array}{c} 10.1 \\ 13.1 \\ 11.0 \\ 12.0 \\ 15.3 \end{array} & \left.\begin{array}{c} 21 \\ 24 \\ 22 \\ 22 \\ 26 \end{array}\right) & \begin{array}{l} \leftarrow \text{Case 1} \\ \leftarrow \text{Case 2} \\ \leftarrow \text{Case 3} \\ \leftarrow \text{Case 4} \\ \leftarrow \text{Case 5} \end{array} \end{array} \tag{1.2}$$

Table 1.2. Age, Height, and Weight for a Sample of Five Children

Child	Age (months)	Weight (pounds)	Height (inches)
1	1	10.1	21
2	2	13.1	24
3	2	11.0	22
4	3	12.0	22
5	4	15.3	26

In statistics, the word *variable* usually refers to the set of measurements for a single characteristic of all individuals. The word *observation* or *case* refers to the set of measurements for all characteristics of an individual. Accordingly, our sample consists of three variables (characteristics): age, weight, and height. It also consists of five observations or cases, one observation per child. Thus, each row in the matrix **X** represents an observation (i.e., a child) and each column represents a variable.

As can be seen from the matrices in (1.1) and (1.2), we should not think of matrices as just arrays of numbers, but rather as data that contain information in an organized, neat, and precise way.

A matrix is referred to by a single boldface letter, such as **A** or **X**, and is treated as a single object. We refer to an individual *element* of a matrix using a double-subscript; the first refers to the row number and the second refers to the column number in which the element is located. Thus, the *ij*th element of **A** is referred to as a_{ij}, which represents the shipping cost from Factory i to City j. For example, $a_{24} = 5$ is the value in the second row and fourth column of **A**, which is the shipping cost from Factory 2 to City 4. Similarly, $x_{21} = 2$ is the age of the second child and $x_{42} = 12.0$ is the weight of the fourth child.

Using subscripts, a generic $n \times m$ matrix **A** can be written as

$$\mathbf{A} = \underbrace{\begin{pmatrix} a_{11} & a_{12} & \cdots & a_{1m} \\ a_{21} & a_{22} & \cdots & a_{2m} \\ \vdots & \vdots & \ddots & \vdots \\ a_{n1} & a_{n2} & \cdots & a_{nm} \end{pmatrix}}_{m \text{ columns}} \Bigg\} \, n \text{ rows},$$

or, more compactly, $\mathbf{A}_{n \times m} = (a_{ij})$, reflecting the fact that **A** is of order $n \times m$ and that a_{ij} is a typical element. If **A** contains more than nine rows and/or nine columns, perhaps a comma could be inserted between the row and column subscripts to eliminate possible confusion. For example, $b_{25,2}$ refers to the element in row 25 and column 2 of **B**, whereas $b_{2,52}$ refers to the element in row 2 and column 52.

Here is a third example of a matrix. Suppose the production costs (per unit) in the three factories mentioned above are \$91, \$93, and \$88, respectively. The production costs can be represented by a matrix **b**, say, as

$$\mathbf{b} = \begin{pmatrix} 91 \\ 93 \\ 88 \end{pmatrix}. \tag{1.3}$$

Here, **b** is a 3×1 matrix. A matrix that has one column is called a *column vector* and a matrix that has one row is called a *row vector*. A matrix of order 1×1 is called a *scalar* (an ordinary number). Thus, for example, the matrix **A** in (1.1) has three row vectors, four column vectors, and twelve scalars.

We shall refer to the order of a vector by the number of its elements; that is, a vector of order $n \times 1$ is referred to as a column vector of order n. Similarly, a vector of order $1 \times n$ is referred to as a row vector of order n. Thus, for example, **b** above is a column vector of order 3. Conventionally, we shall always think of vectors as column vectors, unless otherwise specified.

As mentioned above, the matrix **A** in (1.1) can be described as having four column vectors and three row vectors. Thus, another way of writing a matrix is in terms of its column (or row) vectors. For example, a general matrix **A** of order $n \times m$ can be written in terms of its column vectors, $\mathbf{a}_1, \mathbf{a}_2, \ldots, \mathbf{a}_m$, as

$$\mathbf{A}_{n \times m} = \left(\begin{bmatrix} \vdots \\ \mathbf{a}_1 \\ \vdots \end{bmatrix} \begin{bmatrix} \vdots \\ \mathbf{a}_2 \\ \vdots \end{bmatrix} \cdots \begin{bmatrix} \vdots \\ \mathbf{a}_m \\ \vdots \end{bmatrix} \right) = (\, \mathbf{a}_1 \,\vdots\, \mathbf{a}_2 \,\vdots\, \ldots \,\vdots\, \mathbf{a}_m \,).$$

1.2.2. Scalar Algebra and Matrix Algebra

Algebra has two branches: *scalar* (ordinary) algebra and *matrix* algebra. Scalar algebra deals with scalars, and matrix algebra deals with matrices. Since a scalar is a special case of a matrix, you may think of scalar algebra as a special case of matrix algebra. As such, all the rules of matrix algebra apply to scalar algebra, but the converse is not necessarily true. We shall see examples of this later.

To distinguish among matrices, vectors, and scalars throughout this book, we use boldface uppercase letters to denote matrices, boldface lowercase letters to denote vectors, and italics to denote scalars. Thus, for example,

X and **A** are matrices,
x and **a** are vectors,
x and *a* are scalars.

In a few isolated cases, however, this notational convention may be breached.

1.2.3. Matrix Equality

When we write **A** = **B**, we mean that the matrices **A** and **B** have the same order and that each and every element of **A** is equal to the corresponding element of **B**, that is, $a_{ij} = b_{ij}$ for all values of i and j. In other words, two matrices are said to be equal if and only if they are identical.[2] A definition of *matrix inequality* requires concepts to be presented in Section 11.3.

1.2.4. Matrix Transpose

The transpose of an $n \times m$ matrix **A** (written as $\mathbf{A}^T$ and read as *A-transpose*[3]) is an $m \times n$ matrix obtained by interchanging the rows and columns of **A**; that is, if **A** = (a_{ij}), then $\mathbf{A}^T = (a_{ji})$. For example, the matrix **A** in (1.1) and its transpose are

$$\mathbf{A} = \begin{pmatrix} 5 & 2 & 1 & 4 \\ 2 & 3 & 6 & 5 \\ 7 & 6 & 3 & 2 \end{pmatrix}, \qquad \mathbf{A}^T = \begin{pmatrix} 5 & 2 & 7 \\ 2 & 3 & 6 \\ 1 & 6 & 3 \\ 4 & 5 & 2 \end{pmatrix}.$$

Thus, $\mathbf{A}^T$ is a 4×3 matrix where each row in **A** becomes a column in $\mathbf{A}^T$.

Note, then, that the transpose of a column vector is a row vector; for example, the transpose of the column vector **b** in (1.3) is the row vector

$$\mathbf{b}^T = \begin{pmatrix} 91 & 93 & 88 \end{pmatrix}.$$

Thus, the transpose operator allows us to change a row vector to a column vector and vice versa.

The transpose of a matrix has the following property (other properties are introduced in the following chapters).

---

2 The statement "*x* if and only if *y*" (sometimes abbreviated as *x* iff *y*) is a two-way implication, that is, "*x* implies *y* and also *y* implies *x*." Thus, in this context, if **A** and **B** are identical matrices, then **A** = **B**, and if **A** = **B**, then **A** and **B** are identical.

3 Another popular notation for the transpose is **A**', but since this notation is conventionally used in calculus books to denote the derivative, we use $\mathbf{A}^T$ to denote the transpose of **A**.

Result 1.1. $(\mathbf{A}^{\mathrm{T}})^{\mathrm{T}} = \mathbf{A}$.

This property, which states that the transpose of $\mathbf{A}^{\mathrm{T}}$ is $\mathbf{A}$, is referred to as the *reflexive* property of the transpose.

1.3. Special Types of Matrices

1.3.1. Square Matrices

When the number of rows in a matrix is equal to the number of columns, the matrix is called a *square matrix*. The following matrices are 3×3 square matrices:

$$\mathbf{S} = \begin{pmatrix} 3 & 1 & 4 \\ 1 & 6 & 0 \\ 4 & 0 & 8 \end{pmatrix}, \quad \mathbf{D} = \begin{pmatrix} 5 & 0 & 0 \\ 0 & 0 & 0 \\ 0 & 0 & 1 \end{pmatrix}, \quad \mathbf{I} = \begin{pmatrix} 1 & 0 & 0 \\ 0 & 1 & 0 \\ 0 & 0 & 1 \end{pmatrix}, \quad \mathbf{L} = \begin{pmatrix} 2 & 0 & 0 \\ 8 & 1 & 0 \\ 3 & 0 & 5 \end{pmatrix}. \quad (1.4)$$

Diagonal and off-diagonal elements. A square matrix has two diagonals, a *major diagonal* (sometimes referred to as the *principal diagonal*) and a *minor diagonal*. The major diagonal is the one that runs from the upper-left corner to the lower-right corner of the matrix.[4] The minor diagonal is the one that runs from the upper-right to the lower-left corner. In this book, we make little use of the minor diagonal, so when we say *the diagonal* we mean the major diagonal. The elements on the diagonal of a matrix are naturally called the *diagonal elements* and all other elements are called the *off-diagonal elements*. For example, the diagonal elements of $\mathbf{S}$ in (1.4) are 3, 6, and 8. All other elements are off-diagonal.

1.3.2. Symmetric Matrices

A matrix $\mathbf{A}$ is said to be *symmetric* if $\mathbf{A}^{\mathrm{T}} = \mathbf{A}$ or, in other words, if the elements in the ith row are equal to the corresponding elements in the ith column, for all values of i. Clearly, symmetric matrices have to be square. The matrices $\mathbf{S}$, $\mathbf{D}$, and $\mathbf{I}$ in (1.4) are examples of 3×3 symmetric matrices, whereas the matrix $\mathbf{L}$ is asymmetric.

Symmetric matrices arise naturally and frequently in many practical applications. Consider, for example, the following situation. Suppose there are n people and we wish to record whether or not each person is related to each other person. This information can be arranged neatly and precisely in a matrix like

$$\mathbf{C} = \begin{pmatrix} 1 & 1 & 0 & 0 & 0 \\ 1 & 1 & 0 & 0 & 0 \\ 0 & 0 & 1 & 0 & 0 \\ 0 & 0 & 0 & 1 & 1 \\ 0 & 0 & 0 & 1 & 1 \end{pmatrix} = (c_{ij}).$$

4 One can also define diagonals for any matrix, not necessarily square, but we shall not make use of this definition in this book.

In this case $n = 5$ people. The element c_{ij} indicates whether or not the ith and jth persons are related to each other: a value of 1 indicates that the ith and jth persons are related and a value of 0 means they are not. Clearly, c_{ij} must be equal to c_{ji}, hence the matrix **C** is necessarily symmetric. Note also that $c_{ii} = 1$ for all i (that is, all the diagonal elements are 1), indicating that a person is always related to himself. The matrix **C** indicates that the first two people are related to each other but neither is related to any of the other three. The same is true for the last two people. The third person is not related to any of the other four people. This is an example of a matrix known in the social sciences as the *connectivity* matrix. We shall see many other examples of symmetric matrices in the course of this book.

1.3.3. Skew-Symmetric Matrices

A square matrix **A** is said to be *skew-symmetric* if $\mathbf{A}^{\mathrm{T}} = -\mathbf{A}$, that is, if $a_{ij} = -a_{ji}$. In other words, the elements in the ith row of a skew-symmetric matrix are equal to the negative of the corresponding elements in the ith column, for all values of i. By definition, the diagonal elements of any skew-symmetric matrix must satisfy $a_{ii} = -a_{ii}$. Clearly, the only value that satisfies this condition is $a_{ii} = 0$. It follows then that every diagonal element of a skew-symmetric matrix must be 0. As an example, the matrix

$$\mathbf{K} = \begin{pmatrix} 0 & -1 & 4 \\ 1 & 0 & -2 \\ -4 & 2 & 0 \end{pmatrix} \tag{1.5}$$

is skew-symmetric. In Chapter 2, after matrix addition and subtraction are defined, we shall be able to introduce a systematic way for generating symmetric and skew-symmetric matrices.

1.3.4. Triangular Matrices

When every element below the diagonal of a square matrix is 0 and at least one of the elements above the diagonal is not 0, the matrix is called an *upper triangular matrix*. Similarly, when every element above the diagonal is 0 and at least one of the elements below the diagonal is not 0, the matrix is called a *lower triangular matrix*. For example, the matrices

$$\mathbf{L} = \begin{pmatrix} 2 & 0 & 0 \\ 8 & 0 & 0 \\ 3 & 0 & 5 \end{pmatrix} \quad \text{and} \quad \mathbf{U} = \begin{pmatrix} 1 & 2 & 0 \\ 0 & 1 & 5 \\ 0 & 0 & 1 \end{pmatrix}$$

are triangular; **L** is lower triangular and **U** is upper triangular. When every diagonal element of a triangular matrix is 1, the matrix is called a *unit triangular matrix*. The matrix **U** above is unit upper triangular.

Note that the transpose of a lower (upper) triangular matrix is an upper (lower) triangular matrix. Hence, $\mathbf{L}^{\mathrm{T}}$ is upper triangular and $\mathbf{U}^{\mathrm{T}}$ is lower triangular.

1.3.5. Diagonal Matrices

A *diagonal matrix* is a square matrix in which every off-diagonal element is 0 and at least one of the diagonal elements is not 0. The matrix **D** in (1.4) is an example of a diagonal matrix. If **D** is diagonal, it is usually denoted by $\mathbf{D} = diag(d_{ii})$.

1.3.6. Identity Matrices

An *identity matrix* is a diagonal matrix in which every diagonal element is 1. The matrix **I** in (1.4) is an example of a 3×3 identity matrix. We shall always use **I** to denote an identity matrix. If we wish to make the order of an identity (or any other) matrix explicit, we write it as a subscript. For example, we write the 3×3 identity matrix as $\mathbf{I}_{3 \times 3}$, or simply as $\mathbf{I}_3$. Since $\mathbf{I}_n$ is always square, we shall refer to it as an identity matrix of order n. You may think of the identity matrix as the matrix extension of the scalar number 1. This point is made clearer in Chapter 2.

1.3.7. Unit Vectors

The ith *unit vector* is a vector in which every element is 0 except for the ith element, which is 1. You may visualize the ith unit vector as the ith column of an identity matrix. We denote the ith unit vector by $\mathbf{u}_i$. Thus, for example,

$$\mathbf{u}_1 = \begin{pmatrix} 1 \\ 0 \\ 0 \end{pmatrix} \quad \text{and} \quad \mathbf{u}_2 = \begin{pmatrix} 0 \\ 1 \\ 0 \end{pmatrix}$$

are the first and second unit vectors of order 3, respectively. The order of a unit vector is made clear by the context.

1.3.8. Matrices of Ones

A *matrix of ones*[5] is a matrix in which every element is 1. We use the symbol **J** (or $\mathbf{J}_{n \times m}$ if the order is not clear from the context) to denote the matrix of ones. For example,

$$\mathbf{J}_{2 \times 4} = \begin{pmatrix} 1 & 1 & 1 & 1 \\ 1 & 1 & 1 & 1 \end{pmatrix}$$

is a 2×4 matrix of ones. An important special case of the matrix of ones is the *vector of ones*. We shall use **1** (boldface one), or $\mathbf{1}_n$ if the order is not clear from the context, to denote the vector of ones. Thus, for example,

$$\mathbf{1}_2 = \begin{pmatrix} 1 \\ 1 \end{pmatrix}$$

is a column vector of two ones.

5 The matrix of ones is sometimes called the *unit* matrix, but the name unit matrix might be confused with the identity matrix or with the ith unit vector.

1.3.9. Null Matrices

A *null matrix* is a matrix in which every element is 0. Thus, if at least one element in a matrix is not 0, the matrix is a *nonzero* (or *non-null*) matrix. A null matrix is denoted by $\mathbf{0}$ (boldface zero). Again, we use a subscript if the order is not clear from the context. A null matrix can be thought of as the matrix extension of the scalar 0.

1.3.10. Constant Matrices

A matrix whose elements are all equal is called a *constant matrix*. Thus, both the null matrix and the matrix of ones are constant matrices. An example of a constant row vector is $(2 \quad 2 \quad 2)$.

1.3.11. Partitioned Matrices

A matrix can be partitioned into submatrices. For example, the matrix

$$\mathbf{P} = \begin{pmatrix} 1 & 2 & \vdots & 2 \\ 0 & 1 & \vdots & 4 \\ \cdots & \cdots & \vdots & \cdots \\ 0 & 0 & \vdots & 6 \end{pmatrix} \tag{1.6}$$

is partitioned into four submatrices as indicated by the dotted lines. Of course, a matrix can be partitioned in many alternative ways. For example, $\mathbf{P}$ in (1.6) can be partitioned in a number of different ways, such as

$$\begin{pmatrix} 1 & \vdots & 2 & 2 \\ \cdots & \vdots & \cdots & \cdots \\ 0 & \vdots & 1 & 4 \\ 0 & \vdots & 0 & 6 \end{pmatrix}, \begin{pmatrix} 1 & 2 & \vdots & 2 \\ \cdots & \cdots & \vdots & \cdots \\ 0 & 1 & \vdots & 4 \\ 0 & 0 & \vdots & 6 \end{pmatrix}, \begin{pmatrix} 1 & 2 & 2 \\ \cdots & \cdots & \cdots \\ 0 & 1 & 4 \\ 0 & 0 & 6 \end{pmatrix}, \begin{pmatrix} 1 & 2 & \vdots & 2 \\ 0 & 1 & \vdots & 4 \\ 0 & 0 & \vdots & 6 \end{pmatrix}. \tag{1.7}$$

In this book we shall consider only matrix partitions which consist of adjacent rows and adjacent columns of the original matrix $\mathbf{P}$.

A matrix $\mathbf{P}$ can also be partitioned symbolically as

$$\mathbf{P} = \begin{pmatrix} \mathbf{A} & \mathbf{B} \\ \mathbf{C} & \mathbf{D} \end{pmatrix}, \tag{1.8}$$

where $\mathbf{A}$, $\mathbf{B}$, $\mathbf{C}$, and $\mathbf{D}$ are matrices of appropriate orders. By comparing (1.6) and (1.8), we see that

$$\mathbf{A} = \begin{pmatrix} 1 & 2 \\ 0 & 1 \end{pmatrix}, \quad \mathbf{B} = \begin{pmatrix} 2 \\ 4 \end{pmatrix}, \quad \mathbf{C} = (0 \quad 0), \quad \mathbf{D} = (6). \tag{1.9}$$

A matrix such as $\mathbf{P}$ in (1.8) is called a *partitioned* matrix. When manipulating partitioned matrices, we treat the submatrices as if they are individual elements. For example, we have the following result regarding the transpose of a partitioned matrix.

Result 1.2. The transpose of a partitioned matrix

$$\mathbf{P} = \begin{pmatrix} \mathbf{A} & \mathbf{B} \\ \mathbf{C} & \mathbf{D} \end{pmatrix}$$

is

$$\mathbf{P}^{\mathrm{T}} = \begin{pmatrix} \mathbf{A}^{\mathrm{T}} & \mathbf{C}^{\mathrm{T}} \\ \mathbf{B}^{\mathrm{T}} & \mathbf{D}^{\mathrm{T}} \end{pmatrix}, \qquad (1.10)$$

so that the elements are transposed and the submatrices are also transposed.

To illustrate, suppose we wish to transpose the partitioned matrix **P** in (1.6) in terms of its submatrices defined in (1.9). We first transpose the submatrices and obtain

$$\mathbf{A}^{\mathrm{T}} = \begin{pmatrix} 1 & 0 \\ 2 & 1 \end{pmatrix}, \quad \mathbf{B}^{\mathrm{T}} = (2 \quad 4), \quad \mathbf{C}^{\mathrm{T}} = \begin{pmatrix} 0 \\ 0 \end{pmatrix}, \quad \mathbf{D}^{\mathrm{T}} = (6). \qquad (1.11)$$

We then substitute (1.11) in (1.10) and obtain

$$\mathbf{P}^{\mathrm{T}} = \begin{pmatrix} \mathbf{A}^{\mathrm{T}} & \mathbf{C}^{\mathrm{T}} \\ \mathbf{B}^{\mathrm{T}} & \mathbf{D}^{\mathrm{T}} \end{pmatrix} = \begin{pmatrix} 1 & 0 & \vdots & 0 \\ 2 & 1 & \vdots & 0 \\ \cdots & \cdots & \vdots & \cdots \\ 2 & 4 & \vdots & 6 \end{pmatrix}. \qquad (1.12)$$

Several other properties of partitioned matrices are given in other parts of the book.

Exercises

1.1. Given the following matrices:

$$\mathbf{A} = \begin{pmatrix} 1 & 1 \\ 1 & 1 \\ 1 & 1 \end{pmatrix}, \quad \mathbf{B} = \begin{pmatrix} 1 & 1 & 1 \\ 1 & 1 & 1 \end{pmatrix}, \quad \mathbf{a} = \begin{pmatrix} -1 \\ 0 \\ 1 \end{pmatrix},$$

state whether each of the following statements is true or false:
(a) **A** is square.
(b) **A** is a matrix of ones.
(c) **A** is a unit matrix.
(d) **B** is of order 2×3.
(e) **B** is an identity matrix.
(f) **a** is a column vector.
(g) **A** = **B**.
(h) $\mathbf{A} = \mathbf{B}^{\mathrm{T}}$.

1.2. Find the transpose of each of the following matrices:

$$\mathbf{A} = \begin{pmatrix} 11 & 12 \\ 21 & 22 \\ 31 & 32 \end{pmatrix}, \quad \mathbf{B} = \begin{pmatrix} 0 & -1 & 3 \\ 1 & 5 & 0 \end{pmatrix}, \quad \mathbf{a} = \begin{pmatrix} -1 \\ 0 \\ 1 \end{pmatrix}, \quad \mathbf{C} = (\mathbf{A} \quad \mathbf{B}^{\mathrm{T}} \quad \mathbf{a}).$$

1.3. Given the following matrices:

$$\mathbf{A} = \begin{pmatrix} 1 & 0 & -2 \\ 0 & 0 & -1 \\ -2 & 1 & 4 \end{pmatrix}, \quad \mathbf{B} = \begin{pmatrix} 1 & 0 & 0 \\ 0 & 0 & 0 \\ 0 & 0 & 1 \end{pmatrix}, \quad \mathbf{C} = \begin{pmatrix} 0 & 0 & -2 \\ 0 & 0 & -1 \\ 2 & 1 & 1 \end{pmatrix},$$

state whether each of the following statements is true or false:
(a) **A** is symmetric. (b) **B** is diagonal.
(c) **C** is lower triangular. (d) **C** is skew-symmetric.
(e) **B** contains two null vectors. (f) There is a unit vector in **A**, **B**, and **C**.

1.4. Construct as many partitioned matrices as possible using the following matrices and their transposes: $\mathbf{A}_{2\times 3}$, $\mathbf{b}_{2\times 1}$, $\mathbf{c}_{3\times 1}$, and $\mathbf{d}_{1\times 1}$.

1.5. State whether each of the following statements is true or false:
(a) All identity matrices are square. (b) All symmetric matrices are square.
(c) An identity matrix is also diagonal. (d) All null matrices are symmetric.
(e) It is possible for a diagonal matrix to be an identity matrix.
(f) The transpose of a triangular matrix is a triangular matrix.
(g) The diagonal elements of a skew-symmetric matrix can be positive.

1.6. Estimate the distances between each pair of four cities or towns of your choice. Then construct a matrix **D**, say, containing these inter-city distances. Explain why **D** has to be square, symmetric, and has 0 for each diagonal element.

1.7. There are five rooms in a youth hostel; each room has two beds. The rooms are assigned on a first-come-first-served basis. Two youths will share a room only if they are of the same sex and all the rooms are occupied by at least one person. Two of the first five youths to arrive were men. Accordingly, each is assigned one of the five vacant rooms. Then a group consisting of two men and three women arrived all at once. The receptionist assures them that they all will be accommodated. He started by constructing a square matrix of order 5. Each column represents one of the five rooms. Each row represents one of the just arrived people. The receptionist puts a 1 in the ijth position if the sex of the person in the ith row matches that of the occupant of the jth room; otherwise he puts a zero. The resultant matrix is called the *incidence* matrix because it refers to the incidence or occurrence of a specific event (in this case, whether two people are of the same sex).
(a) Arrange the rooms and individuals the way you wish and construct the corresponding incidence matrix.
(b) After possible rearrangements of the rows and columns of your matrix, show that it can be partitioned as

$$\begin{pmatrix} \mathbf{I}_2 & \mathbf{0}_{2\times 3} \\ \mathbf{0}_{3\times 2} & \mathbf{I}_3 \end{pmatrix}.$$

1.8. Another example of an incidence matrix is encountered when a rumor is spread among a group of people. For simplicity, suppose that a rumor can be spread only among five people. A person can either generate the rumor or hear it from one or more of the other four. The spread of a rumor can be described by an incidence matrix $\mathbf{A} = (a_{ij})$, where $a_{ij} = 1$, if person i passes the rumor to person j, and $a_{ij} = 0$, if person i does not pass the rumor to person j. Two examples of such incidence matrices are

$$\mathbf{A} = \begin{pmatrix} 1 & 0 & 1 & 0 & 1 \\ 0 & 1 & 0 & 1 & 0 \\ 0 & 0 & 1 & 0 & 1 \\ 0 & 0 & 0 & 1 & 0 \\ 0 & 1 & 0 & 0 & 1 \end{pmatrix} \quad \text{and} \quad \mathbf{B} = \begin{pmatrix} 1 & 0 & 1 & 0 & 1 \\ 0 & 1 & 0 & 1 & 0 \\ 1 & 0 & 1 & 0 & 1 \\ 0 & 1 & 0 & 1 & 0 \\ 1 & 0 & 1 & 0 & 1 \end{pmatrix}.$$

Note that the diagonal elements of **A** and **B** are 1 because, presumably, a person passes the rumor to himself. An incidence matrix can also be represented by a *graph* or a *network* or vice versa. In a graph each person is represented by a node. If a rumor can pass from person i to person j, an arrow (known as a *directed link*) is drawn from node i to node j. Otherwise, the two nodes are left unconnected. The resultant graph is known as a network. An example of a network is given in Figure 1.1.

(a) Create the incidence matrix corresponding to the network in Figure 1.1.

(b) Draw the two networks corresponding to the matrices **A** and **B**.

(c) Rearrange the rows and the corresponding columns of **B** so that the resultant matrix is partitioned as

$$\begin{pmatrix} \mathbf{X} & \mathbf{0} \\ \mathbf{0} & \mathbf{Y} \end{pmatrix}.$$

(d) For each of the two matrices **A** and **B**, suppose that a rumor was initially created by the first person. Which of the other four people will eventually hear of the rumor and which will never hear of it?

(e) What would the results be if the rumor was created by the second person rather than the first?

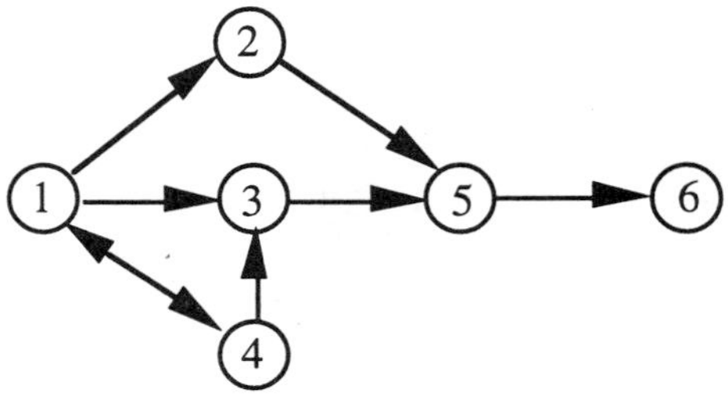

Figure 1.1. An example of a network consisting of six nodes.

2 Some Matrix Calculations

Just like scalars, matrices can be added, subtracted, multiplied, and so forth. Those operations are the subject of this chapter. We postpone matrix division, or inversion, until Chapter 6. We note here that all matrix algebra rules apply to scalar algebra but the converse is not generally true; that is, not all of the scalar algebra rules extend to matrix algebra. Thus, matrix operations usually require more care than scalar operations and we shall point out the scalar algebra rules that do not apply to matrix algebra.

In Section 2.1 we define matrix addition and subtraction. There are three types of matrix multiplications: (a) element-wise or Hadamard product, (b) matrix (inner) product, and (c) Kronecker (outer) product. These types of matrix products are dealt with in Sections 2.2–2.4, respectively.

2.1. Matrix Addition and Subtraction

The sum of two matrices is the matrix containing the sum of the corresponding elements of the two matrices. Similarly, the difference between two matrices is the matrix containing the differences between the corresponding elements of the two matrices. This definition entails that we cannot add or subtract matrices of different orders. If two matrices are of the same order, they are said to be *conformable* or *compatible* for addition and subtraction. Thus, if $\mathbf{A} = (a_{ij})$ and $\mathbf{B} = (b_{ij})$ are of the same order, then we define matrix addition and subtraction as follows:

Matrix Addition: $\mathbf{C} = \mathbf{A} + \mathbf{B}$ implies that $\mathbf{C} = (c_{ij}) = (a_{ij} + b_{ij})$.
Matrix Subtraction: $\mathbf{D} = \mathbf{A} - \mathbf{B}$ implies that $\mathbf{D} = (d_{ij}) = (a_{ij} - b_{ij})$.

Suppose, for example, we wish to compute the total costs of production and shipping of the units from each of the three factories to each of the four cities in Table 1.1. The matrix $\mathbf{A}$ in (1.1), representing the shipping costs, and the vector $\mathbf{b}$ in (1.3), representing production costs, are not conformable for addition, because they are of different orders. To avoid this problem we create another matrix (of the same order as that of $\mathbf{A}$) containing the production costs. Consider the following matrix:

$$\mathbf{P} = \begin{pmatrix} 91 & 91 & 91 & 91 \\ 93 & 93 & 93 & 93 \\ 88 & 88 & 88 & 88 \end{pmatrix}.$$

This matrix has identical columns because the production costs in a given factory are the same regardless of the city to which the goods are shipped. Now, the matrices **A** and **P** are conformable for addition; hence the matrix containing the total production and shipping costs is found by adding **A** and **P**. We obtain

$$\mathbf{T} = \mathbf{A} + \mathbf{P} = \begin{pmatrix} 5 & 2 & 1 & 4 \\ 2 & 3 & 6 & 5 \\ 7 & 6 & 3 & 2 \end{pmatrix} + \begin{pmatrix} 91 & 91 & 91 & 91 \\ 93 & 93 & 93 & 93 \\ 88 & 88 & 88 & 88 \end{pmatrix} = \begin{pmatrix} 96 & 93 & 92 & 95 \\ 95 & 96 & 99 & 98 \\ 95 & 94 & 91 & 90 \end{pmatrix}.$$

Thus, for example, the total production and shipping costs of one unit produced by Factory 2 and shipped to City 4 is 98.

Since scalars are a special case of matrices (of order 1×1), the ordinary rules of scalar addition and subtraction can be viewed as a special case of matrix addition and subtraction. Viewed another way, rules of scalar arithmetic can be extended to matrix addition and subtraction, provided that the matrices are conformable. For example, we have

$$\begin{aligned} &\mathbf{A} + \mathbf{B} = \mathbf{B} + \mathbf{A} \\ &(\mathbf{A} + \mathbf{B}) + \mathbf{C} = \mathbf{A} + (\mathbf{B} + \mathbf{C}) \qquad (2.1) \\ &(\mathbf{A} + \mathbf{B})^{\mathrm{T}} = \mathbf{A}^{\mathrm{T}} + \mathbf{B}^{\mathrm{T}}. \end{aligned}$$

Having defined matrix subtraction, we are in a position to describe a simple way for generating symmetric and skew-symmetric matrices from square matrices, as promised in Chapter 1. This is given below.

Result 2.1. For any *square* matrix **A**,
(a) $\mathbf{A} + \mathbf{A}^{\mathrm{T}}$ is symmetric.
(b) $\mathbf{A} - \mathbf{A}^{\mathrm{T}}$ is skew-symmetric.

2.2. Element-Wise Product

The *element-wise product* of two matrices is defined only if the two matrices are of the same order. The element-wise product of **A** and **B** is a matrix obtained by multiplying each element of **A** by the corresponding element of **B**. Notationally, the element-wise product is indicated by a dot between the two matrices, that is, $\mathbf{A} \cdot \mathbf{B}$. For example, if

$$\mathbf{A} = \begin{pmatrix} 2 & -1 \\ 3 & 0 \end{pmatrix} \quad \text{and} \quad \mathbf{B} = \begin{pmatrix} 1 & 2 \\ 2 & 1 \end{pmatrix},$$

then

$$\mathbf{A} \cdot \mathbf{B} = \begin{pmatrix} 2 & -1 \\ 3 & 0 \end{pmatrix} \cdot \begin{pmatrix} 1 & 2 \\ 2 & 1 \end{pmatrix} = \begin{pmatrix} 2 & -2 \\ 6 & 0 \end{pmatrix}.$$

It can be verified that the element-wise product commutes, that is, $\mathbf{A} \cdot \mathbf{B} = \mathbf{B} \cdot \mathbf{A}$. The element-wise product is also known as the *Hadamard* product.

2.3. Matrix Inner Product

The element-wise product is certainly useful in some applications, but the matrix inner product, which we now define, is more commonly used. For this reason, the matrix inner product is sometimes referred to as the *regular* product. From now on, unless otherwise indicated, *matrix product* will mean the matrix inner (regular) product.

2.3.1. Definitions and Examples

We start first by a definition of the inner product of vectors, then generalize the definition to matrices.

Definition. If **a** and **b** are two vectors, each having n elements, then the *inner product of* **a** and **b** is defined as

$$\sum_{i=1}^{n} a_i b_i ,$$

where $\sum_{i=1}^{n} x_i$ means the sum of x_i as i goes from 1 to n; that is,

$$\sum_{i=1}^{n} x_i = x_1 + x_2 + \cdots + x_n .$$

In words, the inner product of two vectors is a *scalar* obtained by first multiplying the two vectors element-wise ($\mathbf{a} \cdot \mathbf{b}$), then adding up all the resulting products. For example, let

$$\mathbf{a} = \begin{pmatrix} 2 \\ 4 \\ 1 \end{pmatrix} \quad \text{and} \quad \mathbf{b} = \begin{pmatrix} 3 \\ 6 \\ 2 \end{pmatrix}; \tag{2.2}$$

then the element-wise product is

$$\mathbf{a} \cdot \mathbf{b} = \begin{pmatrix} 2 \\ 4 \\ 1 \end{pmatrix} \cdot \begin{pmatrix} 3 \\ 6 \\ 2 \end{pmatrix} = \begin{pmatrix} 6 \\ 24 \\ 2 \end{pmatrix},$$

and the inner product of **a** and **b** is

$$6 + 24 + 2 = 32. \tag{2.3}$$

Note that the inner product of two vectors is always a scalar.

To facilitate the generalization of the above definition to matrices, the inner product of **a** and **b** is written as $\mathbf{a}^T\mathbf{b}$ or $\mathbf{b}^T\mathbf{a}$; that is, the first vector is written as a row vector and the second as a column vector. Note that because the inner product of two vectors is a scalar, we have

$$\mathbf{a}^{\mathrm{T}}\mathbf{b} = \mathbf{b}^{\mathrm{T}}\mathbf{a}. \tag{2.4}$$

The vector inner product may seem unnatural at first, but we shall see soon that the inner product is the most convenient and most common matrix product. As a simple example of its use, suppose we have n numbers $x_1, x_2, \ldots, x_n$, and we wish to compute their sum and the sum of their squares. Thus, we need to compute

$$x_1 + x_2 + \cdots + x_n = \sum_{i=1}^{n} x_i$$

and

$$x_1^2 + x_2^2 + \cdots + x_n^2 = \sum_{i=1}^{n} x_i^2 .$$

To do this using matrix notation, we create two vectors,

$$\mathbf{x} = \begin{pmatrix} x_1 \\ x_2 \\ \vdots \\ x_n \end{pmatrix} \qquad \text{and} \qquad \mathbf{1} = \begin{pmatrix} 1 \\ 1 \\ \vdots \\ 1 \end{pmatrix};$$

the first contains the n numbers and the second is a vector of n ones. The reader can verify that

$$\mathbf{x}^{\mathrm{T}}\mathbf{1} = \mathbf{1}^{\mathrm{T}}\mathbf{x} = \sum_{i=1}^{n} x_i \qquad \text{and} \qquad \mathbf{x}^{\mathrm{T}}\mathbf{x} = \sum_{i=1}^{n} x_i^2 .$$

Thus, $\mathbf{x}^{\mathrm{T}}\mathbf{1}$ or $\mathbf{1}^{\mathrm{T}}\mathbf{x}$ is simply the sum of the elements of $\mathbf{x}$, and $\mathbf{x}^{\mathrm{T}}\mathbf{x}$ is the sum of the squared elements of $\mathbf{x}$. This notation is both elegant and compact, and does not depend on the number of elements in the vectors. We now turn to the products of matrices.

Definition. The product of the two matrices **A** and **B**, denoted by **AB**, is defined as the matrix whose ijth element is equal to the inner product of the ith row of **A** (the matrix on the left) and jth column of **B** (the matrix on the right).

For instance, if $\mathbf{a}_i^{\mathrm{T}}$ is the ith row of **A**, $\mathbf{b}_j$ is the jth column of **B**, and $\mathbf{C} = \mathbf{AB}$, then the ijth element of **C** is the inner product of $\mathbf{a}_i$ and $\mathbf{b}_j$, that is, $c_{ij} = \mathbf{a}_i^{\mathrm{T}}\mathbf{b}_j$. The following diagram illustrates how the elements of the product matrix are obtained as the inner products of the rows of **A** and the columns of **B**:

$$\begin{array}{ccccc} & \mathbf{A}_{n\times m} & \mathbf{B}_{m\times p} & = & \mathbf{C}_{n\times p} & \\ \text{row } i \rightarrow & \begin{pmatrix} \vdots & \vdots & \vdots \\ a_{i1} & \cdots & a_{im} \\ \vdots & \vdots & \vdots \end{pmatrix} & \begin{pmatrix} \cdots & b_{1j} & \cdots \\ \cdots & \vdots & \cdots \\ \cdots & b_{mj} & \cdots \end{pmatrix} & = & \begin{pmatrix} & \vdots & \\ \cdots & c_{ij} & \cdots \\ & \vdots & \end{pmatrix} & \leftarrow \text{row } i \\ & & \uparrow & & \uparrow & \\ & & \text{col. } j & & \text{col. } j & \end{array}$$

Here is an example. Let

$$\mathbf{A} = \begin{pmatrix} 2 & 4 & 1 \\ 1 & -1 & 0 \end{pmatrix}, \quad \mathbf{B} = \begin{pmatrix} 3 & 5 & 0 & 1 \\ 6 & 4 & 2 & 1 \\ 2 & 1 & 2 & 1 \end{pmatrix}, \quad \text{and} \quad \mathbf{C} = \mathbf{AB}; \tag{2.5}$$

then

$$\mathbf{C} = \begin{pmatrix} 2 & 4 & 1 \\ 1 & -1 & 0 \end{pmatrix} \begin{pmatrix} 3 & 5 & 0 & 1 \\ 6 & 4 & 2 & 1 \\ 2 & 1 & 2 & 1 \end{pmatrix} = \begin{pmatrix} 32 & 27 & 10 & 7 \\ -3 & 1 & -2 & 0 \end{pmatrix}.$$

The element

$$c_{11} = \mathbf{a}_1^{\mathrm{T}}\mathbf{b}_1 = 2 \times 3 + 4 \times 6 + 1 \times 2 = 32$$

is the inner product of the first row of **A** and the first column of **B**, the element

$$c_{12} = \mathbf{a}_1^{\mathrm{T}}\mathbf{b}_2 = 2 \times 5 + 4 \times 4 + 1 \times 1 = 27$$

is the inner product of the first row of **A** and the second column of **B**, and so on.

Notice that, in contrast to the vector inner product, the matrices **A** and **B** can be of different orders. For the product **AB** to exist, however, the number of columns in **A** (the matrix on the left) must be equal to the number of rows in **B** (the matrix on the right). So, if **A** is of order $n \times m$ and **B** is of order $q \times p$, then **AB** exists if and only if $m = q$. Furthermore, the order of **AB** (if it exists) is $n \times p$. When the product **AB** exists, that is, when $m = q$, the matrices **A** and **B** are said to be *conformable for multiplication*. In (2.5), for example, **A** is of order 2×3 and **B** is of order 3×4, so that **AB** exists and is of order 2×4.

As mentioned earlier, not all the rules of scalar multiplication extend to matrices. Here are some examples:

1. The product of two scalars always exists, but the product of two matrices does not always exist. For example, for the matrices in (2.5), the product **AB** exists but **BA** does not (because the number of the columns of **B** is not equal to the number of the rows of **A**). For this reason, we need to distinguish between *pre-* and *post-multiplication*. We can describe **AB** as **B** pre-multiplied by **A** or, equivalently, as **A** post-multiplied by **B**.

2. Unlike the scalar product, the matrix product is not in general *commutative*. If $\mathbf{AB} = \mathbf{BA}$, we say that $\mathbf{A}$ and $\mathbf{B}$ *commute*. Even if both $\mathbf{AB}$ and $\mathbf{BA}$ exist, however, these products are not necessarily equal. For example, let

$$\mathbf{A} = \begin{pmatrix} 2 & 0 \\ 1 & 2 \end{pmatrix}, \quad \mathbf{B} = \begin{pmatrix} 1 & 1 \\ 1 & 2 \end{pmatrix};$$

then you can verify that

$$\mathbf{AB} = \begin{pmatrix} 2 & 0 \\ 1 & 2 \end{pmatrix}\begin{pmatrix} 1 & 1 \\ 1 & 2 \end{pmatrix} = \begin{pmatrix} 2 & 2 \\ 3 & 5 \end{pmatrix}$$

and

$$\mathbf{BA} = \begin{pmatrix} 1 & 1 \\ 1 & 2 \end{pmatrix}\begin{pmatrix} 2 & 0 \\ 1 & 2 \end{pmatrix} = \begin{pmatrix} 3 & 2 \\ 4 & 4 \end{pmatrix},$$

hence, in this case, $\mathbf{AB} \neq \mathbf{BA}$. As another example, consider the vectors

$$\mathbf{a} = \begin{pmatrix} 2 \\ 4 \\ 1 \end{pmatrix} \quad \text{and} \quad \mathbf{b} = \begin{pmatrix} 3 \\ 6 \\ 2 \end{pmatrix}$$

as in (2.2). The product $\mathbf{a}^T\mathbf{b}$ was found in (2.3) to be 32; now what is the product $\mathbf{ba}^T$? To find $\mathbf{ba}^T$, we should first ask: does the product exist? Since $\mathbf{b}$ is a 3×1 matrix and $\mathbf{a}^T$ is a 1×3 matrix, then $\mathbf{ba}^T$ is conformable for multiplication and $\mathbf{ba}^T$ exists. You may verify that

$$\mathbf{ba}^T = \begin{pmatrix} 3 \\ 6 \\ 2 \end{pmatrix}(2 \quad 4 \quad 1) = \begin{pmatrix} 6 & 12 & 3 \\ 12 & 24 & 6 \\ 4 & 8 & 2 \end{pmatrix}, \tag{2.6}$$

which is a matrix of order 3×3 as would be expected. We see that $\mathbf{a}^T\mathbf{b} \neq \mathbf{ba}^T$. So, in general, if $\mathbf{a}$ is $m \times 1$ and $\mathbf{b}$ is $m \times 1$, then $\mathbf{a}^T\mathbf{b}$, $\mathbf{b}^T\mathbf{a}$, $\mathbf{ab}^T$, and $\mathbf{ba}^T$ all exist; the first two are scalars and the last two are matrices of order $m \times m$. They are related by $\mathbf{a}^T\mathbf{b} = \mathbf{b}^T\mathbf{a}$ and $(\mathbf{ab}^T)^T = \mathbf{ba}^T$. Note, however, that $\mathbf{ab}^T$ and $\mathbf{ba}^T$ always exist regardless of the orders of $\mathbf{a}$ and $\mathbf{b}$.

2.3.2. Special Types of Matrix Products

We now discuss some special types of matrix products.

Multiplying a matrix by a scalar. The product of a scalar and a matrix is obtained by multiplying each element of the matrix by the scalar, namely,

$$c\mathbf{A} = \begin{pmatrix} ca_{11} & ca_{12} & \cdots & ca_{1m} \\ ca_{21} & ca_{22} & \cdots & ca_{2m} \\ \vdots & \vdots & \ddots & \vdots \\ ca_{n1} & ca_{n2} & \cdots & ca_{nm} \end{pmatrix} = (ca_{ij}).$$

For example, for $c = 3$ and $\mathbf{A} = \begin{pmatrix} 2 & 4 & 1 \\ 1 & -1 & 0 \end{pmatrix}$, we have

$$c\mathbf{A} = 3 \times \begin{pmatrix} 2 & 4 & 1 \\ 1 & -1 & 0 \end{pmatrix} = \begin{pmatrix} 6 & 12 & 3 \\ 3 & -3 & 0 \end{pmatrix}.$$

Note that when the scalar is a positive integer, this product agrees with the definition of matrix addition. Thus, $3\mathbf{A} = \mathbf{A} + \mathbf{A} + \mathbf{A}$, i. e., multiplying $\mathbf{A}$ by a positive integer is equivalent to repeated addition of $\mathbf{A}$ to itself.

Multiplying a matrix by an identity matrix. Pre- or post-multiplying a matrix by the conformable identity matrix leaves the matrix unchanged. Thus, you may think of an identity matrix as the matrix extension of the scalar 1. For example, let

$$\mathbf{A} = \begin{pmatrix} 2 & 4 & 1 \\ 1 & -1 & 0 \end{pmatrix}, \quad \mathbf{I}_2 = \begin{pmatrix} 1 & 0 \\ 0 & 1 \end{pmatrix}, \quad \text{and} \quad \mathbf{I}_3 = \begin{pmatrix} 1 & 0 & 0 \\ 0 & 1 & 0 \\ 0 & 0 & 1 \end{pmatrix},$$

and verify that (a) $\mathbf{AI}_3 = \mathbf{A}$, (b) $\mathbf{I}_2\mathbf{A} = \mathbf{A}$, and (c) neither $\mathbf{AI}_2$ nor $\mathbf{I}_3\mathbf{A}$ exists.

If an identity matrix is the matrix extension of the scalar 1, what is the matrix extension of the scalar 0? It is a *null* (or *zero*) matrix, a matrix in which every element is 0. Thus, for example, $\mathbf{A0} = \mathbf{0}$ and $\mathbf{0A} = \mathbf{0}$, provided that the matrices are conformable for multiplication. Notice that, unless $\mathbf{A}$ is a square matrix, we cannot conclude that $\mathbf{A0} = \mathbf{0A}$ (why?).

Note that the product of two matrices can be a null matrix even though the two matrices are non-null. For example,

$$\begin{pmatrix} 2 & 1 \\ 0 & 0 \end{pmatrix} \begin{pmatrix} -1 & -1 \\ 2 & 2 \end{pmatrix} = \begin{pmatrix} 0 & 0 \\ 0 & 0 \end{pmatrix}.$$

The fact that $\mathbf{AB} = \mathbf{0}$ does not imply that either $\mathbf{A} = \mathbf{0}$ or $\mathbf{B} = \mathbf{0}$ (or both), as it would if $\mathbf{A}$ and $\mathbf{B}$ were scalars.

Multiplying a matrix by a diagonal matrix. Let $\mathbf{D} = (d_{ii})$ be a diagonal matrix. The ith row of the product $\mathbf{DA}$, if it exists, is equal to the ith row of $\mathbf{A}$ multiplied by d_{ii}. Similarly, the jth column of the product $\mathbf{AD}$, if it exists, is equal to the jth column of $\mathbf{A}$ multiplied by d_{jj}. Thus, when we pre-multiply a matrix $\mathbf{A}$ by a conformable diagonal matrix, we work on the rows of $\mathbf{A}$, whereas when we post-multiply a matrix **A** by a conformable diagonal matrix, we work on the columns of **A**. For example, suppose the scores in a mathematics test were recorded in a scale from 0 to 10, but in an economics test they were recorded in a scale of 0 to 50. The scores for three students in both tests are presented in the matrix

$$\mathbf{A} = \begin{pmatrix} 8.5 & 45 \\ 7.0 & 39 \\ 9.0 & 40 \end{pmatrix},$$

where each row of **A** represents the scores of one student in the two tests and each column represents the scores of the three students in one test. Now, suppose we wish to combine the two scores for each student, that is, to compute that student's average for the two test scores. Clearly, the two scores should be measured in the same scale before we add them up. So, let us convert the scores in both tests to a common scale from 0 to 100, say. We can do this by multiplying the scores of the mathematics test by 10 and the scores of the economics test by 2. This task can also be accomplished by creating a diagonal matrix

$$\mathbf{D} = \begin{pmatrix} 10 & 0 \\ 0 & 2 \end{pmatrix};$$

then post-multiply **A** by **D** (since we wish to change the column scale) and obtain

$$\mathbf{AD} = \begin{pmatrix} 8.5 & 45 \\ 7.0 & 39 \\ 9.0 & 40 \end{pmatrix} \begin{pmatrix} 10 & 0 \\ 0 & 2 \end{pmatrix} = \begin{pmatrix} 85 & 90 \\ 70 & 78 \\ 90 & 80 \end{pmatrix} = \mathbf{C}.$$

We see then that the first column of **C** is equal to the first column of **A** multiplied by the first diagonal element of **D**; similarly the second column of **C** is equal to the second column of **A** multiplied by the second diagonal element of **D**. Thus, post-multiplying **A** by a diagonal matrix is equivalent to changing the scales of the columns of **A**. Similarly, pre-multiplying **A** by a diagonal matrix is equivalent to changing the scales of the rows of **A**. The average score for each student can then be computed as

$$\mathbf{G} = \frac{1}{2}\,\mathbf{C}\,\mathbf{1} = \frac{1}{2} \begin{pmatrix} 85 & 90 \\ 70 & 78 \\ 90 & 80 \end{pmatrix} \begin{pmatrix} 1 \\ 1 \end{pmatrix} = \begin{pmatrix} 87.5 \\ 74.0 \\ 85.0 \end{pmatrix},$$

where we multiply **C** by **1** to sum the scores for each student, then divide the sum by the number of scores (2 in this case).

Multiplying a matrix by itself. In some applications we may need to multiply a matrix by itself several times. For example, to find the nth *power* of a matrix **A** we multiply **A** by itself n times and denote the product by $\mathbf{A}^n$. Thus, the second and third powers of **A** are written as $\mathbf{A}^2 = \mathbf{AA}$ and $\mathbf{A}^3 = \mathbf{AAA}$. Clearly, for the second and larger powers of **A** to exist, **A** must be square. It may happen that the powers of a matrix are equal to the matrix itself. Matrices with this property are given a special name as defined below.

Definition. If $\mathbf{A}^2 = \mathbf{A}$, **A** is called an *idempotent* matrix.

It follows from the definition that if $\mathbf{A}^2 = \mathbf{A}$, then $\mathbf{A}^n = \mathbf{A}$, for all positive integers n. The simplest examples of idempotent matrices are the null (square) matrices and the identity matrices. Note also that the numbers 0 and 1 are the only idempotent scalars. Idempotent matrices, other than the null and identity matrices, are encountered in many practical applications and we shall see several examples of them in later chapters.

Multiplying a matrix by a vector. Suppose that **A** is a matrix and **x** is a vector such that the product $\mathbf{y} = \mathbf{A}\mathbf{x}$ exists; then **y** is a vector with the same number of elements as the number of rows in **A**. In this case, one may think of **A** as a *transformation matrix*; it transforms the vector **x** into another vector **y**. Transformations of this kind are called *linear transformations*, and they can be classified even further. For example, if **A** is idempotent, the transformation is a *projection* and if **A** is orthogonal, the transformation is a *rotation*.[1] These and other types of linear transformations are discussed in more detail in Chapter 7. Geometric implications of multiplying a matrix by a vector are presented in Chapter 4.

2.3.3. Some Properties of Matrix Products

We conclude this section with some properties of matrix products.

Result 2.2. Provided that the given matrices are conformable:
(a) $(\mathbf{AB})^T = \mathbf{B}^T\mathbf{A}^T$ (This is known as the *reversal* property of the transpose.)
(b) $(\mathbf{AB})\mathbf{C} = \mathbf{A}(\mathbf{BC})$ (This is known as the *associative* law.)
(c) $\mathbf{A}(\mathbf{B} + \mathbf{C}) = \mathbf{AB} + \mathbf{AC}$ (This is known as the *distributive* law.)
(d) $c\mathbf{A} = \mathbf{A}c$, where c is a scalar.
(e) Both $\mathbf{A}^T\mathbf{A}$ and $\mathbf{A}\mathbf{A}^T$ are symmetric for any matrix **A**.
(f) If **B** is skew-symmetric, then $\mathbf{ABA}^T$ is also skew-symmetric.

Property 2.2(a), in particular, has many useful applications. It stat.es that the transpose of a product of matrices is the product of the transposes of the individual matrices, taken in reverse order.

Result 2.3. Product of partitioned matrices. Let the matrices **X** and **Y** be partitioned as follows:

$$\mathbf{X} = \begin{pmatrix} \mathbf{A} & \mathbf{B} \\ \mathbf{C} & \mathbf{D} \end{pmatrix} \quad \text{and} \quad \mathbf{Y} = \begin{pmatrix} \mathbf{E} & \mathbf{F} \\ \mathbf{G} & \mathbf{H} \end{pmatrix}. \tag{2.7}$$

[1] A square matrix **A** which satisfies $\mathbf{A}^T\mathbf{A} = \mathbf{I}$ is called an orthogonal matrix. Projection and rotation are discussed in detail in Sections 7.3–7.5.

Then, provided that all matrices and submatrices are conformable for the required operations,

$$\mathbf{XY} = \begin{pmatrix} \mathbf{A} & \mathbf{B} \\ \mathbf{C} & \mathbf{D} \end{pmatrix} \begin{pmatrix} \mathbf{E} & \mathbf{F} \\ \mathbf{G} & \mathbf{H} \end{pmatrix} = \begin{pmatrix} \mathbf{AE}+\mathbf{BG} & \mathbf{AF}+\mathbf{BH} \\ \mathbf{CE}+\mathbf{DG} & \mathbf{CF}+\mathbf{DH} \end{pmatrix}. \tag{2.8}$$

Thus, when we multiply two partitioned matrices, we treat the submatrices as if they were individual elements of the matrices.

Having introduced matrix products, we are now in a position to define some special types of vectors and matrices. The vectors

$$\mathbf{a} = \begin{pmatrix} 0.5 \\ -0.5 \\ -0.5 \\ -0.5 \end{pmatrix} \quad \text{and} \quad \mathbf{b} = \begin{pmatrix} 0.5 \\ -0.5 \\ 0.5 \\ 0.5 \end{pmatrix} \tag{2.9}$$

have the property that their inner product is zero; that is,

$$\mathbf{a}^{\mathrm{T}}\mathbf{b} = \begin{pmatrix} 0.5 & -0.5 & -0.5 & -0.5 \end{pmatrix} \begin{pmatrix} 0.5 \\ -0.5 \\ 0.5 \\ 0.5 \end{pmatrix} = 0.$$

Note that this is another example of two matrices (in this case vectors) whose product is the null matrix (in this case a scalar) even though neither one of the two matrices is null.

The matrix

$$\mathbf{A} = \begin{pmatrix} 0.5 & 0.5 & 0.5 & 0.5 \\ -0.5 & -0.5 & 0.5 & 0.5 \\ -0.5 & 0.5 & -0.5 & 0.5 \\ -0.5 & 0.5 & 0.5 & -0.5 \end{pmatrix} \tag{2.10}$$

has the property that

$$\mathbf{A}^{\mathrm{T}}\mathbf{A} = \begin{pmatrix} 0.5 & -0.5 & -0.5 & -0.5 \\ 0.5 & -0.5 & 0.5 & 0.5 \\ 0.5 & 0.5 & -0.5 & 0.5 \\ 0.5 & 0.5 & 0.5 & -0.5 \end{pmatrix} \begin{pmatrix} 0.5 & 0.5 & 0.5 & 0.5 \\ -0.5 & -0.5 & 0.5 & 0.5 \\ -0.5 & 0.5 & -0.5 & 0.5 \\ -0.5 & 0.5 & 0.5 & -0.5 \end{pmatrix}$$

$$= \begin{pmatrix} 1 & 0 & 0 & 0 \\ 0 & 1 & 0 & 0 \\ 0 & 0 & 1 & 0 \\ 0 & 0 & 0 & 1 \end{pmatrix} = \mathbf{I}_4.$$

Vectors and matrices with the above properties are given special names as defined below.

Definitions.

(a) *Normal vector.* If $\mathbf{a}^T\mathbf{a} = 1$, the vector $\mathbf{a}$ is said to be a *normal* vector.

(b) *Orthogonal vectors.* For two vectors $\mathbf{a}$ and $\mathbf{b}$ of the same order, if $\mathbf{a}^T\mathbf{b} = 0$, the two vectors are said to be orthogonal to one another.

(c) *Orthonormal vectors.* For two normal vectors $\mathbf{a}$ and $\mathbf{b}$ of the same order, if $\mathbf{a}^T\mathbf{b} = 0$, the two vectors are said to be orthonormal vectors.

(d) *Orthogonal matrix.* If $\mathbf{A}$ is square and $\mathbf{A}^T\mathbf{A} = \mathbf{I}$, then $\mathbf{A}$ is called an orthogonal matrix.

Note also that if $\mathbf{A}$ is orthogonal, then $\mathbf{A}^T\mathbf{A} = \mathbf{A}\mathbf{A}^T = \mathbf{I}$. Thus, the matrix $\mathbf{A}$ in (2.10) is an orthogonal matrix. You may also verify that all identity matrices are orthogonal matrices. Another example of an orthogonal matrix is

$$\mathbf{A} = \begin{pmatrix} 1/\sqrt{2} & 1/\sqrt{2} \\ 1/\sqrt{2} & -1/\sqrt{2} \end{pmatrix},$$

which contains two orthonormal column (row) vectors. In general, all pairs of vectors (rows or columns) in an orthogonal matrix are orthonormal vectors. However, a matrix may contain orthonormal vectors and yet not be orthogonal. For example, the matrix

$$\mathbf{B} = \begin{pmatrix} 2/3 & -2/3 \\ 1/3 & 2/3 \\ 2/3 & 1/3 \end{pmatrix}$$

contains two orthonormal columns but $\mathbf{B}$ is not orthogonal, because $\mathbf{B}$ is not a square matrix. Note also that although $\mathbf{B}^T\mathbf{B} = \mathbf{I}_2$, $\mathbf{B}\mathbf{B}^T \neq \mathbf{I}_3$. More is said about orthogonal vectors and matrices in other parts of the book.

*2.4. Kronecker Product

A third type of matrix multiplication, which is useful in some applications, is called the *Kronecker* or *direct* product. Let $\mathbf{A}$ be $m \times n$ and $\mathbf{B}$ be $p \times q$; then one form of the Kronecker product of $\mathbf{A}$ and $\mathbf{B}$ is denoted and defined by

$$\mathbf{A}\otimes\mathbf{B} = \begin{pmatrix} a_{11}\mathbf{B} & a_{12}\mathbf{B} & \cdots & a_{1n}\mathbf{B} \\ a_{21}\mathbf{B} & a_{22}\mathbf{B} & \cdots & a_{2n}\mathbf{B} \\ \vdots & \vdots & \ddots & \vdots \\ a_{m1}\mathbf{B} & a_{m2}\mathbf{B} & \cdots & a_{mn}\mathbf{B} \end{pmatrix}. \tag{2.11}$$

Thus, each element of the matrix on the left is multiplied by the matrix on the right. The matrix in (2.11) is partitioned into submatrices. The number of these submatrices is equal to mn, the number of elements in $\mathbf{A}$. Note that $\mathbf{A}\otimes\mathbf{B}$ is a matrix of order $mp \times nq$ and that Kronecker products always exist.

To illustrate, given

$$\mathbf{A} = \begin{pmatrix} 2 & 3 & 1 \\ 1 & -1 & 2 \end{pmatrix} \quad \text{and} \quad \mathbf{B} = \begin{pmatrix} 2 & 0 \\ 0 & 4 \end{pmatrix},$$

then

$$\mathbf{A}\otimes\mathbf{B} = \begin{pmatrix} 2 & 3 & 1 \\ 1 & -1 & 2 \end{pmatrix} \otimes \begin{pmatrix} 2 & 0 \\ 0 & 4 \end{pmatrix} = \begin{pmatrix} 2\begin{pmatrix} 2 & 0 \\ 0 & 4 \end{pmatrix} & 3\begin{pmatrix} 2 & 0 \\ 0 & 4 \end{pmatrix} & 1\begin{pmatrix} 2 & 0 \\ 0 & 4 \end{pmatrix} \\ 1\begin{pmatrix} 2 & 0 \\ 0 & 4 \end{pmatrix} & -1\begin{pmatrix} 2 & 0 \\ 0 & 4 \end{pmatrix} & 2\begin{pmatrix} 2 & 0 \\ 0 & 4 \end{pmatrix} \end{pmatrix}$$

$$= \begin{pmatrix} 4 & 0 & 6 & 0 & 2 & 0 \\ 0 & 8 & 0 & 12 & 0 & 4 \\ 2 & 0 & -2 & 0 & 4 & 0 \\ 0 & 4 & 0 & -4 & 0 & 8 \end{pmatrix}, \tag{2.12}$$

which is a matrix of order 4×6, as would be expected.

Now, let us compute the Kronecker product $\mathbf{B}\otimes\mathbf{A}$. We have

$$\mathbf{B}\otimes\mathbf{A} = \begin{pmatrix} 2 & 0 \\ 0 & 4 \end{pmatrix} \otimes \begin{pmatrix} 2 & 3 & 1 \\ 1 & -1 & 2 \end{pmatrix} = \begin{pmatrix} 2\begin{pmatrix} 2 & 3 & 1 \\ 1 & -1 & 2 \end{pmatrix} & 0\begin{pmatrix} 2 & 3 & 1 \\ 1 & -1 & 2 \end{pmatrix} \\ 0\begin{pmatrix} 2 & 3 & 1 \\ 1 & -1 & 2 \end{pmatrix} & 4\begin{pmatrix} 2 & 3 & 1 \\ 1 & -1 & 2 \end{pmatrix} \end{pmatrix}$$

$$= \begin{pmatrix} 4 & 6 & 2 & 0 & 0 & 0 \\ 2 & -2 & 4 & 0 & 0 & 0 \\ 0 & 0 & 0 & 8 & 12 & 4 \\ 0 & 0 & 0 & 4 & -4 & 8 \end{pmatrix}. \tag{2.13}$$

It can be seen from (2.12) and (2.13) that

$$\mathbf{A}\otimes\mathbf{B} \neq \mathbf{B}\otimes\mathbf{A}. \tag{2.14}$$

However, by inspection of (2.12) and (2.13), we see that the elements of $\mathbf{A}\otimes\mathbf{B}$ are the same as the elements of $\mathbf{B}\otimes\mathbf{A}$ except that they are in different positions. In fact, either of the two matrices in (2.12) and (2.13) can be obtained by permuting the rows and columns of the other. Nevertheless, in some cases we may have $\mathbf{A}\otimes\mathbf{B} = \mathbf{B}\otimes\mathbf{A}$. Examples of such cases are given below.

Kronecker product of a matrix and a scalar. The Kronecker product of a scalar c and a matrix $\mathbf{A}$ is obtained by multiplying the scalar by each element of the matrix. Thus, the Kronecker product of a scalar and a matrix is the same as the regular product of the scalar and the matrix. Therefore, we can write

$$c\otimes\mathbf{A} = \mathbf{A}\otimes c = c\mathbf{A} = \mathbf{A}c = (ca_{ij}). \tag{2.15}$$

Kronecker product of two vectors. Let us, for example, find $\mathbf{b}\otimes\mathbf{a}^{\mathrm{T}}$, where

$$\mathbf{a} = \begin{pmatrix} 2 \\ 4 \\ 1 \end{pmatrix} \quad \text{and} \quad \mathbf{b} = \begin{pmatrix} 3 \\ 6 \\ 2 \end{pmatrix}$$

as in (2.2). We have

$$\begin{aligned}
\mathbf{b}\otimes\mathbf{a}^{\mathrm{T}} &= \begin{pmatrix} 3 \\ 6 \\ 2 \end{pmatrix} \otimes \begin{pmatrix} 2 & 4 & 1 \end{pmatrix} \\
&= \begin{pmatrix} 3(2 & 4 & 1) \\ 6(2 & 4 & 1) \\ 2(2 & 4 & 1) \end{pmatrix} = \begin{pmatrix} 6 & 12 & 3 \\ 12 & 24 & 6 \\ 4 & 8 & 2 \end{pmatrix},
\end{aligned}$$

which is the same as the product $\mathbf{b}\mathbf{a}^{\mathrm{T}}$ computed in (2.6). Thus, the product of $\mathbf{b}$ (a column vector on the left) and $\mathbf{a}^{\mathrm{T}}$ (a row vector on the right), usually called the *outer product* of $\mathbf{a}$ and $\mathbf{b}$, is the same as the Kronecker product $\mathbf{b}\otimes\mathbf{a}^{\mathrm{T}}$. So, in general, for any two vectors $\mathbf{a}$ and $\mathbf{b}$, we have

$$\mathbf{b}\otimes\mathbf{a}^{\mathrm{T}} = \mathbf{a}^{\mathrm{T}}\otimes\mathbf{b} = \mathbf{b}\mathbf{a}^{\mathrm{T}}. \tag{2.16}$$

Kronecker product of a matrix and a diagonal matrix. Notice that (2.12) is the Kronecker product of a matrix $\mathbf{A}$ on the left and a diagonal matrix $\mathbf{B}$ on the right. This product can actually be obtained by

$$\mathbf{A}\otimes\mathbf{B} = (\mathbf{A}\otimes\mathbf{I})(\mathbf{I}\otimes\mathbf{B}), \tag{2.17}$$

where the first **I** is an identity matrix of the same order as that of **A** and the second **I** is an identity matrix of the same order as that of **B**. Thus, in this case, the Kronecker product $\mathbf{A}\otimes\mathbf{B}$ is equal to the (regular) product of $\mathbf{A}\otimes\mathbf{I}$ and $\mathbf{I}\otimes\mathbf{B}$. Recall from (2.14), however, that $\mathbf{A}\otimes\mathbf{B}$ does not have to be equal to $\mathbf{B}\otimes\mathbf{A}$.

The Kronecker product has many other interesting properties, some of which are listed below, while others are given in later parts of the book.

Result 2.4. Kronecker products have the following properties:
(a) $(\mathbf{A}\otimes\mathbf{B})^{\mathrm{T}} = \mathbf{A}^{\mathrm{T}}\otimes\mathbf{B}^{\mathrm{T}}$.
(b) $(\mathbf{A}\otimes\mathbf{B})(\mathbf{C}\otimes\mathbf{D}) = \mathbf{AC}\otimes\mathbf{BD}$, provided that the regular products **AC** and **BD** exist.
(c) Let **P** be a partitioned matrix of the form $\mathbf{P} = (\mathbf{A} : \mathbf{B})$; then for any matrix **C**,

$$\mathbf{P}\otimes\mathbf{C} = (\mathbf{A} : \mathbf{B})\otimes\mathbf{C} = (\mathbf{A}\otimes\mathbf{C} : \mathbf{B}\otimes\mathbf{C}), \tag{2.18}$$

but $\mathbf{C}\otimes(\mathbf{A} : \mathbf{B})$ is not necessarily equal to $(\mathbf{C}\otimes\mathbf{A} : \mathbf{C}\otimes\mathbf{B})$.

Notice the difference between the transpose of the regular product (Result 2.2(a)) and the transpose of the Kronecker product (Result 2.4(a)). The transpose of the latter is the product of the transpose of the individual matrices, but the matrices stay in the same order.

Exercises

2.1. Using the matrices $\mathbf{A} = \begin{pmatrix} 2 & -1 \\ 3 & 5 \end{pmatrix}$, $\mathbf{B} = \begin{pmatrix} 1 & 0 \\ -2 & 4 \end{pmatrix}$, and $\mathbf{C} = \begin{pmatrix} 6 & 3 \\ 2 & 1 \end{pmatrix}$, verify the addition and subtraction rules specified in (2.1).

2.2. Verify Result 2.1 using $\mathbf{A} = \begin{pmatrix} 2 & -1 \\ 0 & 4 \end{pmatrix}$.

2.3. For any two vectors **a** and **b** of the same order:
(a) Show that the inner product of **a** and **b** can be expressed as $(\mathbf{a}\cdot\mathbf{b})^{\mathrm{T}}\mathbf{1}$, where $\mathbf{a}\cdot\mathbf{b}$ is the element-wise product of **a** and **b**, and **1** is a conformable vector of ones.
(b) Show that $\mathbf{a}^{\mathrm{T}}\mathbf{b} = \mathbf{b}^{\mathrm{T}}\mathbf{a}$, as claimed in (2.4).

2.4. Given the following matrices:

$$\mathbf{A} = \begin{pmatrix} 2 & 3 \\ 0 & -5 \\ 1 & 4 \end{pmatrix}, \quad \mathbf{B} = \begin{pmatrix} 1 & -2 & 0 \\ 1 & 3 & 4 \end{pmatrix}, \quad \mathbf{x} = \begin{pmatrix} -1 \\ 0 \\ 1 \end{pmatrix}, \quad \mathbf{y} = \begin{pmatrix} 2 \\ 1 \end{pmatrix},$$

Indicate whether each of the following expressions is defined and calculate it where possible:
(a) $\mathbf{A} + \mathbf{B}^{\mathrm{T}}$ (b) $\mathbf{B} - \mathbf{A}^{\mathrm{T}}$ (c) $\mathbf{B}\cdot\mathbf{A}^{\mathrm{T}}$ (d) $\mathbf{B}\cdot\mathbf{x}$

(e) $\mathbf{x}^T\mathbf{y}$ and $\mathbf{y}^T\mathbf{x}$ (f) $\mathbf{x}^T\mathbf{x}$ (g) $\mathbf{A} - \mathbf{x}\mathbf{y}^T$ (h) $\mathbf{Bx} + \mathbf{y}$
(i) $2\mathbf{A}$ (j) $\mathbf{A}^T\mathbf{A}$ (k) $\mathbf{A}\mathbf{A}^T$ (l) $(\mathbf{A} + \mathbf{B}^T)^T\mathbf{x}$

2.5. In each of the following cases, indicate whether the matrices **A** and **B** commute:

(a) $\mathbf{A} = \begin{pmatrix} 2 & -1 \\ 0 & 5 \end{pmatrix}$ and $\mathbf{B} = \begin{pmatrix} 2 & 0 \\ 0 & 2 \end{pmatrix}$. (b) $\mathbf{A} = \begin{pmatrix} 0 & -2 \\ 0 & 4 \end{pmatrix}$ and $\mathbf{B} = \begin{pmatrix} 2 & 3 \\ 0 & 0 \end{pmatrix}$.

(c) $\mathbf{A} = \begin{pmatrix} 2 & -1 \\ 0 & 4 \end{pmatrix}$ and $\mathbf{B} = \begin{pmatrix} 2 & 0 \\ 0 & 4 \end{pmatrix}$. (d) $\mathbf{A} = \begin{pmatrix} 2 & -1 \\ 0 & 5 \end{pmatrix}$ and $\mathbf{B} = \begin{pmatrix} 4 & 1 \\ 0 & 1 \end{pmatrix}$.

(e) $\mathbf{A}_{2\times 3}$ and $\mathbf{B}_{3\times 2}$. (f) $\mathbf{A}_{3\times 3}$ and $\mathbf{B} = \mathbf{I}_3$.

2.6. Let $\mathbf{A} = \begin{pmatrix} 2 & -1 \\ 0 & 5 \end{pmatrix}$ and $\mathbf{B} = \begin{pmatrix} b_{11} & b_{12} \\ b_{21} & b_{22} \end{pmatrix}$.

(a) What conditions must the elements of **B** satisfy so that **A** and **B** commute?
(b) Find a matrix **B** such that **A** and **B** commute.
(c) Suppose that **A** and **B** commute; is **B** unique?

2.7. Given the following matrices:

$$\mathbf{A} = \frac{1}{2}\begin{pmatrix} 1 & \sqrt{3} \\ 0 & 0 \\ -\sqrt{3} & 1 \end{pmatrix}, \mathbf{B} = \frac{1}{2}\begin{pmatrix} \sqrt{3} & 1 \\ -1 & \sqrt{3} \end{pmatrix}, \mathbf{x} = \frac{1}{\sqrt{2}}\begin{pmatrix} -1 \\ 0 \\ 1 \end{pmatrix}, \mathbf{y} = \begin{pmatrix} 1 \\ 2 \\ -1 \end{pmatrix}, \mathbf{z} = \begin{pmatrix} 1/3 \\ 1/3 \\ 1/3 \end{pmatrix},$$

compute:
(a) $\mathbf{A}^T\mathbf{A}$ (b) $\mathbf{A}\mathbf{A}^T$ (c) $\mathbf{B}^T\mathbf{B}$ (d) $\mathbf{AB}$
(e) $\mathbf{x}^T\mathbf{x}$ (f) $\mathbf{z}^T\mathbf{z}$ (g) $\mathbf{x}^T\mathbf{z}$ (h) $\mathbf{x}^T\mathbf{y}$
Then state whether and why each of the following statements is true or false:
(i) **A** and **B** are orthogonal matrices. (j) **A** is orthogonal but **B** is not.
(k) **A** is orthogonal to **B**. (l) **x** and **z** are normal vectors.
(m) **x** and **z** are orthogonal vectors. (n) **x** and **y** are orthonormal vectors.

2.8. For any angle θ, show that the matrix

$$\mathbf{A} = \begin{pmatrix} \cos\theta & \sin\theta \\ -\sin\theta & \cos\theta \end{pmatrix}$$

is orthogonal. [Hint: Use the trigonometric identity $(\sin\theta)^2 + (\cos\theta)^2 = 1$.]

2.9. Let $\mathbf{I}_n$ be the identity matrix of order n. For any $n \times m$ matrix **A**, show that:
(a) The rows of **A** are orthogonal if and only if $\mathbf{A}\mathbf{A}^T$ is diagonal.
(b) The columns of **A** are orthogonal if and only if $\mathbf{A}^T\mathbf{A}$ is diagonal.
(c) The rows of **A** are orthonormal if and only if $\mathbf{A}\mathbf{A}^T = \mathbf{I}_n$.
(d) The columns of **A** are orthonormal if and only if $\mathbf{A}^T\mathbf{A} = \mathbf{I}_m$.

2.10. Let $\mathbf{A} = \begin{pmatrix} a & -b \\ b & a \end{pmatrix}$.

(a) What are the conditions under which $\mathbf{A}$ is an orthogonal matrix?
(b) Select two values of a and b so that $\mathbf{A}$ is an orthogonal matrix.

2.11. (a) Show that if $\mathbf{B}$ is skew-symmetric and $\mathbf{A}^T\mathbf{BA}$ exists, then $\mathbf{A}^T\mathbf{BA}$ is skew-symmetric.
(b) Verify Result 2.2 using the matrices:

$$\mathbf{A} = \begin{pmatrix} 2 & 1 \\ 0 & 2 \\ 1 & 0 \end{pmatrix}, \quad \mathbf{B} = \begin{pmatrix} 0 & -2 \\ 2 & 0 \end{pmatrix}, \quad \mathbf{C} = \begin{pmatrix} 3 & 1 \\ 1 & 2 \end{pmatrix}, \quad c = 2.$$

2.12. (a) Find any matrix $\mathbf{A}$, other than the $\mathbf{I}$ and $\mathbf{0}$ matrices, such that $\mathbf{A}^2 = \mathbf{A}$.
(b) Using your matrix in (a), find a non-null matrix $\mathbf{B}$ such that $\mathbf{AB} = \mathbf{0}$.
(c) Find any non-null matrix $\mathbf{A}$ such that $\mathbf{A}^2 = \mathbf{0}$.

2.13. Given $\mathbf{A} = \begin{pmatrix} 1 & 2 & 5 \\ 2 & 4 & 10 \\ -1 & -2 & -5 \end{pmatrix}$, $\mathbf{B} = \begin{pmatrix} 1 & 0 \\ 4 & -1 \end{pmatrix}$, and $\mathbf{C} = \begin{pmatrix} 3 & -2 \\ 3 & -2 \end{pmatrix}$, show that:

(a) $\mathbf{A}^2 = \mathbf{0}$. (b) $\mathbf{B}^2 = \mathbf{I}$. (c) $\mathbf{C}^2 = \mathbf{C}$.

2.14. Let $\mathbf{A}$ and $\mathbf{B}$ be two matrices. Determine which properties they must satisfy to ensure that $(\mathbf{A} + \mathbf{B})^2$ is defined and equal to:
(a) $\mathbf{A}^2 + \mathbf{B}^2 + \mathbf{AB} + \mathbf{BA}$ (b) $\mathbf{A}^2 + \mathbf{B}^2 + 2\mathbf{AB}$ (c) $\mathbf{A}^2 + \mathbf{B}^2$

2.15. Let $\mathbf{x}$ and $\mathbf{1}$ be vectors of order n. Answer true or false:
(a) $\mathbf{1x}^T$ is a square matrix.
(b) Both $\mathbf{x}^T\mathbf{1}$ and $\mathbf{1}^T\mathbf{x}$ are scalars.
(c) The matrix $\mathbf{1x}^T$ has identical rows.
(d) The matrix $\mathbf{1x}^T$ is symmetric.
(e) $\mathbf{11}^T$ is a matrix of ones.
(f) $\mathbf{1}^T\mathbf{1} = n$.
(g) The average element of $\mathbf{x}$ is $n^{-1}\mathbf{x}^T\mathbf{1}$.
(h) $(n^{-1}\mathbf{x}^T\mathbf{1})^T = n^{-1}\mathbf{1}^T\mathbf{x}$.

2.16. Let $a_{ij} = (i - j)^2$ be the ijth element of a 2×3 matrix $\mathbf{A}$.
(a) Write $\mathbf{A}$ explicitly.
(b) Compute $\mathbf{A} \cdot \mathbf{A}$ (the Hadamard product of $\mathbf{A}$ and itself).
(c) Find a matrix $\mathbf{B}$ such that $\mathbf{A} = \mathbf{B} \cdot \mathbf{B}$.
(d) Either prove that $\mathbf{B}$ is unique or find another example of $\mathbf{B}$.

2.17. When fully stocked, an automatic vending machine takes 50 cans of juice, 25 cans of mineral water, and 25 pints of milk. At the end of every day, the owner of the machine refills it to its capacity and empties the coin box. The drinks are sold for 75 cents per can of juice, 50 cents per can of mineral water, and 75 cents per pint of milk. During a given day, 40 cans of soda, 20 cans of mineral water, and 22 pints of milk were sold.

(a) Create a vector **c** containing the machine capacity of each drink, a vector **s** containing the units sold for each drink, and a vector **p** containing the price per each drink. Use these vectors, and others if necessary, to answer the remaining questions.
(b) What is the total value (selling price) of the drinks in the machine when fully stocked?
(c) Compute the vector representing the inventory left in the machine at the end of the day.
(d) How much money does the owner expect to find in the coin box at the end of the day?

2.18. Let $\mathbf{x}$ be a vector containing the first n positive integers and s_n be their sum. Also, let $\mathbf{1}$ be the vector of n ones and $a_{ij} = 10i + j$ be the ijth element of an $m \times n$ matrix $\mathbf{A}$.
(a) Show that $s_n = n(n+1)/2$. (b) What is the ijth element of $\mathbf{A}^{\mathrm{T}}$?
(c) Construct an example of a matrix $\mathbf{A}$ as defined above with $m = n = 3$.
(d) Show that the sum of all elements of $\mathbf{A}$ in (c) is $\sum_{i=1}^{3} \sum_{j=1}^{3} a_{ij} = 198$.
(e) For $m = n$, show that $\mathbf{A} = 10\mathbf{x}\mathbf{1}^{\mathrm{T}} + \mathbf{1}\mathbf{x}^{\mathrm{T}}$.
(f) For $m = n$, show that $\sum_{i=1}^{n} \sum_{j=1}^{n} a_{ij} = \mathbf{1}^{\mathrm{T}}\mathbf{A}\mathbf{1} = 11n\, s_n$.
(g) Find a general expression similar to that in (e) for any m and n. [Hint: Use two vectors, one of order m and the other of order n.]
(h) Using your expression in (g), show that $\sum_{i=1}^{m} \sum_{j=1}^{n} a_{ij} = 10n\, s_m + m\, s_n$.

2.19. Let $\mathbf{x}$ and $\mathbf{1}$ be vectors of order n. Let $\bar{x} = n^{-1}\mathbf{x}^{\mathrm{T}}\mathbf{1}$ be the arithmetic mean of the elements of $\mathbf{x}$.
(a) For $\mathbf{x} = (-1 \quad 0 \quad 2 \quad 3 \quad 1)^{\mathrm{T}}$, compute and interpret the vector $\mathbf{x} - \bar{x}\mathbf{1}$.
(b) For $\mathbf{x}$ in (a), compute and interpret the quantity $(\mathbf{x} - \bar{x}\mathbf{1})^{\mathrm{T}}(\mathbf{x} - \bar{x}\mathbf{1})$.
(c) Show that $(\mathbf{x} - \bar{x}\mathbf{1})^{\mathrm{T}}(\mathbf{x} - \bar{x}\mathbf{1}) = \mathbf{x}^{\mathrm{T}}\mathbf{x} - n\bar{x}^2$, for any $\mathbf{x}$.
(d) The variance of the elements in $\mathbf{x}$ is defined as $s^2 = \sum_{i=1}^{n}(x_i - \bar{x})^2 / (n-1)$.

Show that $s^2 = (n-1)^{-1}(\mathbf{x}^{\mathrm{T}}\mathbf{x} - n\bar{x}^2)$.

2.20. A network consists of n nodes. Any pair of nodes can be either connected or unconnected. Let $\mathbf{A}$ be a matrix whose ijth element is 1 if the nodes i and j are connected, and 0 otherwise. A node i is assumed to be unconnected with itself ($a_{ii} = 0$). The matrix $\mathbf{A}$ is called an *adjacency* matrix because it contains information about whether any two nodes in the network are adjacent (connected) to each other. For example, the following diagram shows how the adjacency matrix is obtained from a given network of four nodes:

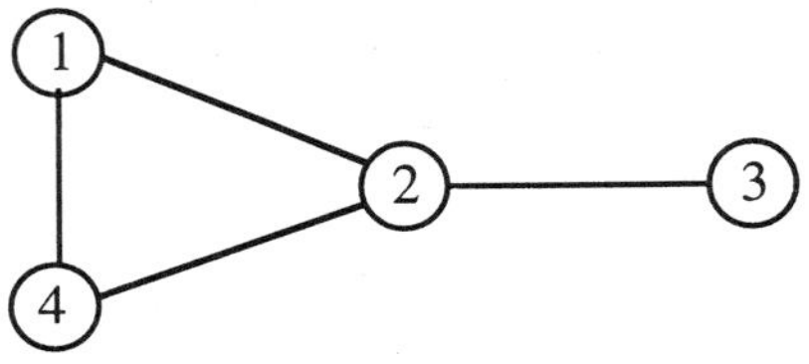

$$\Rightarrow \quad \mathbf{A} = \begin{pmatrix} 0 & 1 & 0 & 1 \\ 1 & 0 & 1 & 1 \\ 0 & 1 & 0 & 0 \\ 1 & 1 & 0 & 0 \end{pmatrix}$$

Two questions that usually arise in practice are: (a) How many paths are there between every pair of nodes? and (b) What is the length (the number of links) of each of these paths (if any)? For example, in the above network, there are two paths between nodes 1 and 3. They are: (1 - 2 - 3) of length 2 and (1 - 4 - 2 - 3) of length 3. The answers to the above questions are provided by the following interesting result:[2] The number of paths of length r between nodes i and j is given by the ijth element of $\mathbf{A}^r$, the rth power of the adjacency matrix. Compute $\mathbf{A}^2$ and $\mathbf{A}^3$, then verify, for example, that there are only one path of length 2 and one path of length 3 joining the nodes 1 and 3.

2.21. Let $\mathbf{X} = \begin{pmatrix} 2 & 0 \\ 0 & 2 \\ 0 & 3 \end{pmatrix} = \begin{pmatrix} \mathbf{X}_1 \\ \mathbf{X}_2 \end{pmatrix}$ and $\mathbf{Z} = \begin{pmatrix} 3 & -1 & 1 \\ -2 & 0 & 0 \\ 2 & 0 & 0 \end{pmatrix} = \begin{pmatrix} \mathbf{Z}_{11} & \mathbf{Z}_{12} \\ \mathbf{Z}_{21} & \mathbf{Z}_{22} \end{pmatrix}$.

(a) Find the product **ZX** by direct multiplication.
(b) Find the product **ZX** using Result 2.3 after any appropriate partitioning of the two matrices as indicated above.

2.22. Given the following partitioned matrices:

$$\mathbf{X} = \begin{pmatrix} \mathbf{X}_1 \\ \mathbf{X}_2 \end{pmatrix}, \quad \mathbf{y} = \begin{pmatrix} \mathbf{y}_1 \\ \mathbf{y}_2 \end{pmatrix}, \quad \mathbf{Z} = \begin{pmatrix} \mathbf{Z}_{11} & \mathbf{Z}_{12} \\ \mathbf{Z}_{21} & \mathbf{Z}_{22} \end{pmatrix},$$

for each of the following cases, state the conditions (on the orders of the partitions) for the product to exist and find the indicated product:

(a) $\mathbf{X}^T\mathbf{X}$ (b) $\mathbf{X}\mathbf{X}^T$ (c) $\mathbf{X}^T\mathbf{y}$ (d) $\mathbf{y}^T\mathbf{y}$ (e) $\mathbf{Z}^T\mathbf{y}$ (f) $\mathbf{Z}^T\mathbf{Z}$

2.23. Given $\mathbf{A} = \begin{pmatrix} 2 & -1 \\ -1 & 4 \end{pmatrix}$, $\mathbf{B} = \begin{pmatrix} 1 & 0 \\ 0 & 1 \end{pmatrix}$, $\mathbf{c} = \begin{pmatrix} 1 \\ 3 \end{pmatrix}$, $d = 2$, compute:

(a) $\mathbf{A} \otimes d$ (b) $\mathbf{A} \otimes \mathbf{c}$ (c) $\mathbf{c} \otimes \mathbf{A}$ (d) $\mathbf{A} \otimes \mathbf{B}$ (e) $\mathbf{A} \otimes \mathbf{A}$ (f) $\mathbf{c} \otimes \mathbf{c}$

2.24. Verify Result 2.4 using the matrices

$$\mathbf{A} = \begin{pmatrix} 2 & -1 \\ -1 & 4 \end{pmatrix}, \quad \mathbf{B} = \begin{pmatrix} 1 & 0 \\ 0 & 1 \end{pmatrix}, \quad \mathbf{C} = \begin{pmatrix} 0 & 2 \\ 1 & 0 \end{pmatrix}, \quad \mathbf{D} = \begin{pmatrix} 2 & 0 \\ 1 & 1 \end{pmatrix}.$$

---

2 For a proof of this and other results and for applications of graph and network theory, see, for example, Castillo, Gutiérrez, and Hadi (1996).

3 Linear Dependence and Independence

3.1. Introduction

The notions of linear dependence and independence play a pivotal role in many applications of matrix algebra, such as solving systems of simultaneous linear equations (Section 8.1). In this chapter we define these concepts and discuss a method for determining whether a set of vectors is linearly dependent or independent. First we need the following definition.

Definition. A sum of scalar multiples of a set of vectors is called a *linear combination* of the vectors. Thus, if $a_1, a_2, \ldots, a_n$ are scalars and $\mathbf{v}_1, \mathbf{v}_2, \ldots, \mathbf{v}_n$ are vectors, then

$$\mathbf{c} = a_1\,\mathbf{v}_1 + a_2\,\mathbf{v}_2 + \cdots + a_n\,\mathbf{v}_n \tag{3.1}$$

is a linear combination of the vectors. Using matrix notation, the reader can verify that (3.1) can be written as

$$\mathbf{c} = \mathbf{V}\mathbf{a}, \tag{3.2}$$

where

$$\mathbf{V} = \begin{pmatrix}\mathbf{v}_1 & \mathbf{v}_2 & \cdots & \mathbf{v}_n\end{pmatrix} = \begin{pmatrix} v_{11} & v_{12} & \cdots & v_{1n} \\ v_{21} & v_{22} & \cdots & v_{2n} \\ \vdots & \vdots & \ddots & \vdots \\ v_{m1} & v_{m2} & \cdots & v_{mn} \end{pmatrix} \quad \text{and} \quad \mathbf{a} = \begin{pmatrix} a_1 \\ a_2 \\ \vdots \\ a_n \end{pmatrix}.$$

As an example, consider the vectors

$$\mathbf{v}_1 = \begin{pmatrix}2\\1\\0\end{pmatrix}, \quad \mathbf{v}_2 = \begin{pmatrix}1\\0\\3\end{pmatrix}, \quad \mathbf{v}_3 = \begin{pmatrix}4\\1\\6\end{pmatrix} \tag{3.3}$$

and let $a_1 = 2$, $a_2 = -2$, and $a_3 = 0$; then

$$\mathbf{c} = 2\mathbf{v}_1 + (-2)\mathbf{v}_2 + 0\,\mathbf{v}_3$$

$$= 2\begin{pmatrix} 2 \\ 1 \\ 0 \end{pmatrix} + (-2)\begin{pmatrix} 1 \\ 0 \\ 3 \end{pmatrix} + 0\begin{pmatrix} 4 \\ 1 \\ 6 \end{pmatrix} = \begin{pmatrix} 2 \\ 2 \\ -6 \end{pmatrix}$$

is a linear combination of the three vectors. Since $\mathbf{c} = 2(\mathbf{v}_1 - \mathbf{v}_2)$, this particular linear combination can be interpreted as twice the difference between the first and second vectors.

Clearly, a linear combination of nonzero vectors can be nonzero only if the scalars are not all 0. Even if the scalars are not all 0, however, a linear combination can still be $\mathbf{0}$ (a zero vector[1]). For example, using the vectors in (3.3) and setting $a_1 = 1$, $a_2 = 2$, and $a_3 = -1$, the reader can verify that $\mathbf{v}_1 + 2\mathbf{v}_2 - \mathbf{v}_3 = \mathbf{0}$. If the scalars are all 0, the linear combination is called a *trivial* linear combination; otherwise it is called *nontrivial.*

The notion of linear combination facilitates the definition of linear dependence or independence of a set of vectors. We first define the linear dependence or independence of a set of two vectors, then generalize the definition to the case where the set contains more than two vectors.

3.2. Linear Dependence of Two Vectors

Definition. Two nonzero vectors $\mathbf{v}_1$ and $\mathbf{v}_2$ are *linearly dependent* if one vector is a nonzero scalar multiple of the other, that is, if

$$\mathbf{v}_1 = a\mathbf{v}_2, \tag{3.4}$$

where $a \neq 0$. If the vectors are not linearly dependent, they are said to be *linearly independent.*

For example, let us determine whether the vectors

$$\mathbf{x} = \begin{pmatrix} -4 \\ 4 \end{pmatrix}, \quad \mathbf{y} = \begin{pmatrix} -2 \\ 2 \end{pmatrix}, \quad \mathbf{z} = \begin{pmatrix} 4 \\ 4 \end{pmatrix}, \quad \mathbf{w} = \begin{pmatrix} 2 \\ 6 \end{pmatrix}, \tag{3.5}$$

are pair-wise linearly dependent. For the pair $\mathbf{x}$ and $\mathbf{y}$, we see that $\mathbf{x} = 2\mathbf{y}$. Thus $\mathbf{x}$ and $\mathbf{y}$ are linearly dependent. Let us now consider the pair $\mathbf{x}$ and $\mathbf{z}$. In this case, we cannot write one vector as a multiple of the other. Therefore, $\mathbf{x}$ and $\mathbf{z}$ are linearly independent. The reader can verify that all of the remaining pairs, $\{(\mathbf{x}, \mathbf{w}), (\mathbf{y}, \mathbf{z}), (\mathbf{y}, \mathbf{w}), (\mathbf{z}, \mathbf{w})\}$, are pair-wise linearly independent.

To facilitate the generalization of the definition of linear dependence to the case where the set contains more than two vectors, we offer an alternative definition. Let a_1 and a_2 be two scalars such that $a_1 \neq 0$ and let $a = -a_2/a_1$. Equation (3.4) can then be written as $\mathbf{v}_1 = (-a_2/a_1)\mathbf{v}_2$ or $a_1\mathbf{v}_1 = -a_2\mathbf{v}_2$ or $a_1\mathbf{v}_1 + a_2\mathbf{v}_2 = \mathbf{0}$. This last equation indicates that a nontrivial linear combination of two dependent vectors can

[1] Recall that a nonzero vector is a vector in which at least one element is not 0.

produce a zero vector. For example, the two vectors **x** and **y** in (3.5) can be seen to satisfy $\mathbf{x} - 2\mathbf{y} = \mathbf{0}$, in which case $a_1 = 1$ and $a_2 = -2$. This leads to the following definition of linear dependence or independence.

Definition. Two vectors $\mathbf{v}_1$ and $\mathbf{v}_2$ are said to be *linearly dependent* if a nontrivial linear combination of the two vectors is equal to a zero vector, that is, if there exist scalars a_1 and a_2, at least one of which is not 0, such that

$$a_1\,\mathbf{v}_1 + a_2\,\mathbf{v}_2 = \mathbf{0}. \tag{3.6}$$

If the vectors are not linearly dependent, they are said to be *linearly independent.*

Using matrix notation, (3.6) can be written as $\mathbf{Va} = \mathbf{0}$, where

$$\mathbf{V} = \begin{pmatrix} \mathbf{v}_1 & \mathbf{v}_2 \end{pmatrix} = \begin{pmatrix} v_{11} & v_{12} \\ v_{21} & v_{22} \\ \vdots & \vdots \\ v_{m1} & v_{m2} \end{pmatrix} \quad \text{and} \quad \mathbf{a} = \begin{pmatrix} a_1 \\ a_2 \end{pmatrix}.$$

Taking, for example, the vectors **x** and **y** in (3.5), we have

$$\mathbf{Va} = \begin{pmatrix} -4 & -2 \\ 4 & 2 \end{pmatrix} \begin{pmatrix} 1 \\ -2 \end{pmatrix} = \begin{pmatrix} 0 \\ 0 \end{pmatrix}.$$

3.3. Linear Dependence of a Set of Vectors

The definition of dependence and independence of two vectors can be extended to three or more vectors as follows.

Definition. Let $\mathbf{v}_1, \mathbf{v}_2, \ldots, \mathbf{v}_n$ be a set of n vectors each of which is of order m. Then the set of vectors is said to be *linearly dependent* if there exist scalars a_1, a_2, ..., a_n, at least one of which is not 0, such that

$$a_1\,\mathbf{v}_1 + a_2\,\mathbf{v}_2 + \cdots + a_n\,\mathbf{v}_n = \mathbf{0}, \tag{3.7}$$

or, equivalently,

$$\mathbf{Va} = \mathbf{0}. \tag{3.8}$$

If the vectors are not linearly dependent, they are said to be *linearly independent.* Clearly, when $n = 2$, (3.7) reduces to (3.6). Now, without loss of generality, suppose $a_1 \neq 0$. Dividing both sides of (3.7) by a_1 and arranging terms, we obtain

$$\mathbf{v}_1 = -\,(a_2 / a_1)\,\mathbf{v}_2 - \cdots - (a_n / a_1)\,\mathbf{v}_n,$$

which implies that when a set of vectors is linearly dependent, at least one of its vectors can be written as a linear combination of the others. In other words, at least one vector linearly depends on the others. For example, you may verify that any pair of the vectors

$$\mathbf{v}_1 = \begin{pmatrix} 2 \\ 1 \\ 0 \end{pmatrix}, \qquad \mathbf{v}_2 = \begin{pmatrix} 1 \\ 0 \\ 3 \end{pmatrix}, \qquad \mathbf{v}_3 = \begin{pmatrix} 4 \\ 1 \\ 6 \end{pmatrix}$$

is linearly independent, yet the set of three vectors is linearly dependent. This is because no one vector can be expressed as a multiple of another, yet each of the vectors can be expressed as a linear combination of the other two. For example, one can show that $\mathbf{v}_3 = \mathbf{v}_1 + 2\,\mathbf{v}_2$, which implies $\mathbf{v}_1 + 2\,\mathbf{v}_2 - \mathbf{v}_3 = \mathbf{0}$. Put in the notation of (3.8), we have

$$\mathbf{V} = \begin{pmatrix} 2 & 1 & 4 \\ 1 & 0 & 1 \\ 0 & 3 & 6 \end{pmatrix}, \qquad \mathbf{a} = \begin{pmatrix} 1 \\ 2 \\ -1 \end{pmatrix},$$

so that

$$\mathbf{Va} = \begin{pmatrix} 2 & 1 & 4 \\ 1 & 0 & 1 \\ 0 & 3 & 6 \end{pmatrix} \begin{pmatrix} 1 \\ 2 \\ -1 \end{pmatrix} = \begin{pmatrix} 0 \\ 0 \\ 0 \end{pmatrix}.$$

We now address the following question: Given a set of vectors, how can we determine whether the vectors are linearly dependent or independent? To answer this question we need the following definition.

Definition. A matrix is said to be in a *row-echelon*[2] form if the number of leading zeros (if any) in each successive row is increasing until all the elements in the last rows are zeros (if any).

Accordingly, a matrix in a row-echelon form might look like this:

$$\begin{pmatrix} 0 & y & x & x & x & x & x \\ 0 & 0 & 0 & y & x & x & x \\ 0 & 0 & 0 & 0 & y & x & x \\ 0 & 0 & 0 & 0 & 0 & 0 & 0 \\ 0 & 0 & 0 & 0 & 0 & 0 & 0 \end{pmatrix}, \tag{3.9}$$

where the y represents nonzero numbers and x represents numbers that may or may not be 0. The number of leading zeros in the five rows, $\{1, 3, 4, 7, 7\}$, respectively, increases until we encounter the last two rows of zeros. Consider, for example, the following matrices:

---

[2] There is an analogous definition for a column-echelon form, a brief discussion of which is given in Section 3.7.

$$\mathbf{A} = \begin{pmatrix} 2 & 1 & 4 \\ 0 & 1 & 2 \\ 0 & 0 & 0 \end{pmatrix}, \quad \mathbf{B} = \begin{pmatrix} 0 & 2 & 3 & 0 \\ 0 & 0 & 1 & 5 \\ 0 & 0 & 0 & 0 \\ 0 & 0 & 0 & 0 \end{pmatrix}, \quad \mathbf{C} = \begin{pmatrix} 0 & 2 & -4 & 6 \\ 1 & 2 & 3 & 0 \\ 0 & -1 & 2 & 4 \\ -2 & -4 & -9 & 1 \end{pmatrix}. \tag{3.10}$$

It can be seen that **A** is in a row-echelon form because the number of leading zeros in the three rows, $\{0, 1, 3\}$, is increasing. The matrix **B** is also in a row-echelon form because the number of leading zeros in the four rows, $\{1, 2, 4, 4\}$, is increasing until we encounter last two rows of zeros. The matrix **C** is not in a row-echelon form, because the number of leading zeros in the four rows, $\{1, 0, 1, 0\}$, is not increasing.

Can any matrix be transformed into a row-echelon form? Is a row-echelon form of a matrix unique? How does a row-echelon form of a matrix determine whether the vectors are dependent or independent? The answers to these questions are given in the following result.

Result 3.1.

(a) Any matrix **A** can be reduced to a row-echelon form by an appropriate sequence of operations known as *elementary row operations.*

(b) A row-echelon form of a matrix is not unique.

(c) All row-echelon forms of a matrix have the same number of nonzero rows.

(d) The rows of **A** are linearly dependent if and only if a row-echelon form of **A** contains at least one all-zero row; otherwise, the rows are linearly independent.

(e) The number of nonzero rows in a row-echelon form of **A** is equal to the number of nonzero rows in a row-echelon form of $\mathbf{A}^T$.

(f) The number of linearly independent vectors (rows or columns) in **A** is equal to the number of nonzero rows in a row-echelon form of **A**.

Result 3.1 has many important implications. First, if the matrix is not already in a row-echelon form, it can always be reduced to a row-echelon form. In Section 3.5 we discuss the steps required to transform any matrix to an equivalent row-echelon form, and in the appendix to this chapter we give some illustrative examples.

Second, a matrix can have many row-echelon forms because a particular row-echelon form of a matrix depends on the particular sequence of elementary row operations performed on the matrix. Notwithstanding the non-uniqueness of row-echelon forms, Result 3.1(c)–(f) indicate that a matrix, its transpose, and all of their echelon forms have the same number of linearly independent vectors. This leads to the third and most important implication of Result 3.1, namely, to determine whether a set of vectors is linearly dependent or independent, one can execute the following steps:

Step 1. Create a matrix **A** containing the set of vectors as rows.

Step 2. If **A** is not already in a row-echelon form, find any row-echelon form of **A** (as outlined in Section 3.5).

Step 3. Examine the obtained row-echelon form of **A** and, using Result 3.1(d), determine whether the set is linearly dependent or independent.

To illustrate, let us now examine some of the above matrices and determine whether they contain linearly dependent or independent vectors. The 5×7 matrix in (3.9) is already in a row-echelon form. This row-echelon form contains three nonzero rows and two rows of zeros. Thus, by Result 3.1(d), the five row vectors in this matrix are linearly dependent. We also know from Result 3.1(f) that, of the five vectors, there cannot be more than three linearly independent vectors.

For the matrices in (3.10), **A** and **B** are already in a row-echelon form. Both **A** and **B** contain rows of zeros, hence each matrix contains a set of linearly dependent row vectors. Furthermore, the maximum number of linearly independent vectors in each matrix is two, because each contains only two nonzero rows.

The matrix **C** is not in a row-echelon form. A row-echelon form of **C** is found in the appendix to this chapter (see (3.19)) to be

$$\begin{pmatrix} 1 & 2 & 3 & 0 \\ 0 & 2 & -4 & 6 \\ 0 & 0 & -3 & 1 \\ 0 & 0 & 0 & 7 \end{pmatrix},$$

which contains no zero rows. Therefore, all four rows in **C** are linearly independent.

We end this section with two results, one regarding the maximum number of linearly independent vectors in a set of vectors and the other regarding the representation of any vector as a linear combination of the unit vectors.

Result 3.2. Let **A** be a set of n vectors each of which is of order m. Then the maximum number of linearly independent vectors in **A** is less than or equal to $\min(m, n)$.

For example, we can speak of the matrix in (3.9) as a set containing seven column-vectors or as a set containing five row-vectors. Of the seven column vectors, each of order 5, there can be at most five linearly independent columns ($\min(7, 5) = 5$), and of the five row vectors, each of order 7, there can be at most five linearly independent rows ($\min(5, 7) = 5$). Of course, the actual number of linearly independent vectors (rows or columns) in this matrix is three.

Result 3.3. Any vector **v** of order m can be expressed as a linear combination of m unit vectors of the same order;[3] that is,

$$\mathbf{v} = v_1\, \mathbf{u}_1 + v_2\, \mathbf{u}_2 + \cdots + v_n\, \mathbf{u}_m,$$

where v_i is the ith element of **v** and $\mathbf{u}_i$ is the ith unit vector of order m.

Note that since $\mathbf{u}_1, \mathbf{u}_2, ..., \mathbf{u}_m$ are the columns of the identity matrix of order m, Result 3.3 is another representation of the simple fact that

$$(\mathbf{u}_1 \quad \mathbf{u}_2 \quad \cdots \quad \mathbf{u}_m)\mathbf{v} = \mathbf{I}\mathbf{v} = \mathbf{v},$$

---

3 Recall, from Chapter 1, that the ith unit vector of order n is the ith column of the identity matrix of order n.

where **I** is the identity matrix of order m. For example, the vector

$$\mathbf{v} = \begin{pmatrix} 3 \\ 2 \\ -1 \end{pmatrix}$$

can be written as

$$\mathbf{v} = 3\begin{pmatrix} 1 \\ 0 \\ 0 \end{pmatrix} + 2\begin{pmatrix} 0 \\ 1 \\ 0 \end{pmatrix} - 1\begin{pmatrix} 0 \\ 0 \\ 1 \end{pmatrix} = \begin{pmatrix} 1 & 0 & 0 \\ 0 & 1 & 0 \\ 0 & 0 & 1 \end{pmatrix}\begin{pmatrix} 3 \\ 2 \\ -1 \end{pmatrix} = \begin{pmatrix} 3 \\ 2 \\ -1 \end{pmatrix}.$$

In the remainder of this chapter we discuss some topics related to the concepts of linear dependence and independence of a set of vectors. Specifically, in Section 3.4 we define the rank of a matrix. In Section 3.5 we discuss elementary row operations, which are useful, among other things, in computing a row-echelon form of a matrix. This discussion leads to a definition of a type of matrix known as *elementary* matrices (Section 3.6). Finally, in Section 3.7 we introduce the concept of elementary column operations, the analogue of elementary row operations.

3.4. Rank of a Matrix

The notion of linear independence facilitates the following definition.

Definition. The number of linearly independent rows (or columns) of a matrix **A** is called the *rank* of **A** and is denoted by r(**A**).
Consider, for example, the matrices

$$\mathbf{A} = \begin{pmatrix} 2 & 1 \\ 1 & 2 \end{pmatrix} \quad \text{and} \quad \mathbf{B} = \begin{pmatrix} 2 & 1 \\ 4 & 2 \end{pmatrix}.$$

You may verify that **A** has two linearly independent columns (rows), hence r(**A**) = 2, and that the two columns (rows) of **B** are linearly dependent, hence r(**B**) = 1.

Note that the rank is defined for any matrix, whether it is square or rectangular. We have seen in Result 3.1 that the number of linearly independent rows (or columns) is equal to the number of nonzero rows in a row-echelon form of the matrix. Therefore, the rank of a matrix is unique and can be determined from any of its row-echelon forms. For example, since **A** and **B** in (3.10) are already in row-echelon form and each has two nonzero rows, r(**A**) = r(**B**) = 2. The matrix **C** in (3.10) is not in a row-echelon form, but one of its row-echelon forms is $\mathbf{C}_4$, which is found in the appendix to this chapter (see (3.19)). Since $\mathbf{C}_4$ has four nonzero rows, r(**C**) = r($\mathbf{C}_4$) = 4.

The rank of a matrix has the following properties:

Result 3.4. For any matrix **A**, the number of linearly independent rows is equal to the number of linearly independent columns, hence r(**A**) is unique.

Result 3.5. For any matrix $\mathbf{A}$, the following is true:

(a) $r(\mathbf{0}) = 0$, where $\mathbf{0}$ is the null matrix.

(b) If $\mathbf{A}$ is of order $n \times m$, then $r(\mathbf{A}) \leq \min(n, m)$.

(c) $r(\mathbf{A}^T) = r(\mathbf{A})$.

(d) $r(\mathbf{A}\mathbf{A}^T) = r(\mathbf{A}^T\mathbf{A}) = r(\mathbf{A})$.

(e) $r(\mathbf{AB}) \leq \min(r(\mathbf{A}), r(\mathbf{B}))$.

(f) $r(\mathbf{A} + \mathbf{B}) \leq r(\mathbf{A}) + r(\mathbf{B})$.

(g) If $\mathbf{A}$ is $m \times n$ of rank $k > 0$, then there are k linearly independent rows and k linearly independent columns of $\mathbf{A}$.

(h) *Rank factorization*: If $\mathbf{A}$ is $n \times m$ of rank $k > 0$, then $\mathbf{A}$ can be expressed as

$$\mathbf{A} = \mathbf{P}\begin{pmatrix} \mathbf{I}_k & \mathbf{0} \\ \mathbf{0} & \mathbf{0} \end{pmatrix}\mathbf{Q},$$

for some square matrices $\mathbf{P}$ and $\mathbf{Q}$, where $\mathbf{I}_k$ is the identity matrix of order k, and the orders of the null matrices are implicit.

(i) $r(\mathbf{A}\otimes\mathbf{B}) = r(\mathbf{A})\, r(\mathbf{B})$.

Part (a) of the above result states that the rank of all null matrices is 0. The reason for this is that any null matrix is already in a row-echelon form and there are no nonzero rows in a null matrix. Part (b) states that the rank of a matrix cannot be larger than its number of rows or number of columns, whichever is smaller. If $r(\mathbf{A})$ is equal to the number of columns in $\mathbf{A}$, the matrix $\mathbf{A}$ is said to be of *full-column* rank. Similarly, if $r(\mathbf{A})$ is equal to the number of the rows, the matrix is said to be of *full-row* rank. If $r(\mathbf{A}) < \min(n, m)$, the matrix $\mathbf{A}$ is called *rank-deficient*.

Part (c), which is a consequence of Part (b), indicates that transposition does not change the rank. Note that Part (d) states that $\mathbf{A}\mathbf{A}^T$ and $\mathbf{A}^T\mathbf{A}$ have the same rank as $\mathbf{A}$, although they are not necessarily of the same order. Part (e) states that the rank of a product of matrices cannot be larger than the smallest of the individual ranks, whereas Part (f) indicates that the rank of a sum cannot exceed the sum of the ranks.

Part (g) is self-explanatory. The rank factorization in Part (h) is a very useful result; it states that, for a given matrix $\mathbf{A}$ of rank k, there exist two matrices $\mathbf{P}$ and $\mathbf{Q}$ such that $\mathbf{A}$ can be factored into a product of the three matrices indicated in Result 3.5(h). We shall return to this property in Section 6.3.6 and make a remark about how the matrices $\mathbf{P}$ and $\mathbf{Q}$ are found and show how $\mathbf{P}$ and $\mathbf{Q}$ are useful in obtaining another matrix factorization known as the full-rank factorization and in obtaining a generalized inverse of a matrix.

Finally, Part (i) asserts that the rank of a Kronecker product is the product of the individual ranks.

3.5. Elementary Row Operations

As stated in Result 3.1(a), any matrix $\mathbf{A}$ can be transformed into a row-echelon form. In practice, calculating a row-echelon form of a matrix or determining whether a set of vectors is linearly independent is usually done by computer. Nevertheless, we briefly present an algorithm for computing a row-echelon form of a matrix because this algorithm is (1) both simple and constructive; (2) useful for other matrix calculations such as finding the *rank* or *determinant* of a matrix (Chapter 5), matrix inversion (Chapter 6), and solving systems of simultaneous linear equations (Section 8.1); and (3) useful for theoretical and conceptual manipulations of matrices. The algorithm consists of a series of matrix operations known as *elementary row operations*, which are defined as follows.

Definition. Given a matrix **A**, one can perform the following three types of elementary row operations on **A**:

Type 1. Interchanging any two rows.
Type 2. Multiplying a row by a nonzero scalar.
Type 3. Adding to a row a nonzero multiple of another row.

Thus, by performing these operations, systematically and repeatedly, on the rows of a given matrix, the matrix can always be transformed into a row-echelon form. Illustrative examples of these operations are given in the appendix to this chapter.

3.6. Elementary Matrices

As stated above, elementary row operations can be used to reduce a matrix to a row-echelon form. These operations can be carried out by pre-multiplying the matrix by a sequence of matrices known as elementary matrices.

Definition. An *elementary matrix* is a matrix obtained from the identity matrix after applying one elementary row operation.

Two implications of this definition are that elementary matrices are necessarily square and that there are three types of elementary matrices (one for each type of elementary row operation). The three types of elementary matrices are:

Type 1. A matrix obtained by interchanging any two rows of an identity matrix.
Type 2. A matrix obtained by multiplying any row of an identity matrix by a nonzero scalar.[4]
Type 3. A matrix obtained by adding to a row of an identity matrix a nonzero multiple of another row.

Thus, to obtain an elementary matrix, one first creates an identity matrix and then performs one type of elementary row operation on it. The appendix to this chapter gives some illustrative examples.

An important property of elementary matrices is that they can be used to perform elementary row operations. These operations can be carried out on a matrix by pre-multiplying it by appropriate elementary matrices as stated in the following result.

Result 3.6. A row-echelon form of any matrix **A** can always be obtained by pre-multiplying **A** by an appropriate sequence of elementary matrices.

---

4 Note that this is equivalent to multiplying the corresponding diagonal element of the identity matrix by the scalar.

Thus, for example, suppose that r elementary row operations are required to reduce **A** to a row-echelon form **B**; then **B** can be written as

$$\mathbf{E}_r\mathbf{E}_{r-1} \cdots \mathbf{E}_2\mathbf{E}_1\mathbf{A} = \mathbf{B}, \tag{3.11}$$

where $\mathbf{E}_1, \mathbf{E}_2, \ldots, \mathbf{E}_r$ are elementary matrices of appropriate type and order. Note that since we are working on the rows of **A**, we pre-multiply **A** by the sequence of elementary matrices. Again, illustrative examples are given in the appendix to this chapter.

3.7. Elementary Column Operations

So far we have been dealing with row operations and row-echelon forms as opposed to *column* operations and *column-echelon* forms. Column operations are the same as row operations but they are performed on the columns of a matrix. Column-echelon forms are obtained by carrying out elementary operations on the columns of the matrix. Column-elementary matrices can also be obtained by carrying out the operations on the columns of an identity matrix. In this case, however, since we are working on the columns of **A**, we *post-multiply* **A** by a sequence of column-elementary matrices. Accordingly, if **B** is a column-echelon form of **A**, then **B** can be expressed as

$$\mathbf{A}\mathbf{E}_1\mathbf{E}_2 \cdots \mathbf{E}_{c-1}\mathbf{E}_c = \mathbf{B}, \tag{3.12}$$

where c is the number of elementary column operations required to reduce **A** to a column-echelon form and $\mathbf{E}_1, \mathbf{E}_2, \ldots, \mathbf{E}_c$ are elementary matrices of appropriate type and order. Equation (3.12) is the analogue of (3.11).

3.8. Permutation Matrices

In this section we introduce a new type of matrix, known as permutation matrices, which are related to elementary operations of the first type. Suppose we have an $m \times n$ matrix **A** and we wish to perform a sequence of r row operations of Type 1 (row interchanges) and a sequence of c column operations of Type 1 (column interchanges). Again, we can perform this task either directly on the matrix **A**, or by applying a sequence of elementary matrices of Type 1. The latter is performed in two steps. We first create an appropriate sequence of row-elementary matrices of Type 1,

$$\mathbf{E}_r\mathbf{E}_{r-1} \cdots \mathbf{E}_2\mathbf{E}_1 = \mathbf{R}, \tag{3.13}$$

then pre-multiply **A** by **R** and obtain **RA**. The product **RA** is the matrix **A** after applying the desired sequence of row interchanges. Second, we create an appropriate sequence of column-elementary matrices of Type 1,

$$\mathbf{E}_1\mathbf{E}_2 \cdots \mathbf{E}_{c-1}\mathbf{E}_c = \mathbf{C}, \tag{3.14}$$

then post-multiply **RA** by **C** and obtain **RAC**. The product **RAC** is the matrix **RA** after applying the desired sequence of column interchanges. The resultant matrix, **RAC**, is the matrix **A** after applying the desired sequence of row and column interchanges. The matrices **R** and **C** are called *permutation* matrices because their effect on **A** is simply to permute its rows and columns. As indicated here, a permutation matrix can be thought of as a product of elementary matrices of Type 1.

Note that the r row-elementary matrices in (3.13) are of order $m \times m$, whereas the c column-elementary matrices in (3.14) are of order $n \times n$. Of course it makes no difference whether we first interchange the rows and then the columns or vice versa, because by the associative law of Result 2.2(b), we have $(\mathbf{RA})\mathbf{C} = \mathbf{R}(\mathbf{AC})$.

We end this section by stating some properties that apply to all elementary matrices whether they are row- or column-elementary.

Result 3.7. Let **A** be an elementary matrix of order $n \times n$. Then:
(a) **A** has n linearly independent rows and n linearly independent columns.
(b) The rank of **A** is n.

Part (b) follows immediately from Part (a).

Appendix: Reduction to Row-Echelon Form

This appendix provides illustrative examples of the various types of row operations and the corresponding elementary matrices required to transform a given matrix to a row-echelon form. For this purpose we use some of the matrices given in this chapter.

Elementary Row Operations

Consider first the matrix **C** from (3.10):

$$\mathbf{C} = \begin{pmatrix} 0 & 2 & -4 & 6 \\ 1 & 2 & 3 & 0 \\ 0 & -1 & 2 & 4 \\ -2 & -4 & -9 & 1 \end{pmatrix}. \tag{3.15}$$

In order to reduce this matrix to a row-echelon form, we start with the first row. We see that $c_{11} = 0$ but not all the elements below it are zero. We therefore interchange row 1 with either row 2 or row 4 so that the element c_{11} becomes nonzero. If we interchange row 1 with row 2, we obtain

$$\mathbf{C}_1 = \begin{pmatrix} 1 & 2 & 3 & 0 \\ 0 & 2 & -4 & 6 \\ 0 & -1 & 2 & 4 \\ -2 & -4 & -9 & 1 \end{pmatrix} \begin{matrix} \mathbf{r}_2 \\ \mathbf{r}_1 \\ \mathbf{r}_3 \\ \mathbf{r}_4 \end{matrix}. \tag{3.16}$$

Thus, $\mathbf{C}_1$ is obtained from **C** by interchanging two rows of **C**, that is, by applying one row operation of Type 1.

To show how the rows of a matrix are obtained from the previous matrix after applying one or more row operations, we use the shorthand notation written next to each row on the right of the new matrix. Here the notation $\mathbf{r}_2$ next to the first row and $\mathbf{r}_1$ next to the second row of $\mathbf{C}_1$ indicates that the first and second rows of the previous matrix ($\mathbf{C}$ in this case) have been interchanged. The notation also indicates that rows 3 and 4 of $\mathbf{C}_1$ are the same as rows 3 and 4 of $\mathbf{C}$.

Now that the element in the (1,1) position of $\mathbf{C}_1$ is nonzero, we reduce every element below it to 0. An examination of the first column of $\mathbf{C}_1$ indicates that we need only replace -2 in the (4,1) position by 0. This can be done by performing a row operation of Type 3, namely, multiplying the first row by 2 and adding the result to the fourth row. We then obtain

$$\mathbf{C}_2 = \begin{pmatrix} 1 & 2 & 3 & 0 \\ 0 & 2 & -4 & 6 \\ 0 & -1 & 2 & 4 \\ 0 & 0 & -3 & 1 \end{pmatrix} \begin{matrix} \mathbf{r}_1 \\ \mathbf{r}_2 \\ \mathbf{r}_3 \\ \mathbf{r}_4 + 2\mathbf{r}_1 \end{matrix} \tag{3.17}$$

Again, the notation indicates that the first three rows of $\mathbf{C}_1$ and $\mathbf{C}_2$ are the same and the fourth row of $\mathbf{C}_2$ is obtained by adding to the fourth row of $\mathbf{C}_1$ twice the first row.

Every element below the (1,1) position is now 0, so we move to the second row. The element in the (2,2) position is nonzero and we need to replace every nonzero element below it by 0. To replace the -1 in the (3,2) position by 0, we multiply row 2 by 0.5 and add the result to row 3. We obtain

$$\mathbf{C}_3 = \begin{pmatrix} 1 & 2 & 3 & 0 \\ 0 & 2 & -4 & 6 \\ 0 & 0 & 0 & 7 \\ 0 & 0 & -3 & 1 \end{pmatrix} \begin{matrix} \mathbf{r}_1 \\ \mathbf{r}_2 \\ \mathbf{r}_3 + 0.5\mathbf{r}_2 \\ \mathbf{r}_4 \end{matrix}. \tag{3.18}$$

Note that each time we perform an operation, we must be careful not to replace any of the leading zeros by a nonzero number. For example, the element in the (3,2) position in $\mathbf{C}_2$ could be replaced by 0 by multiplying row 1 by 0.5 and adding the result to row 3; but this would replace the 0 in the (3,1) position by a nonzero number.

Now that every element below the (2,2) position is 0, we move to the third row. The element in the (3,3) position is 0, but the element below it is nonzero. So, we can interchange row 3 and row 4 and obtain

$$\mathbf{C}_4 = \begin{pmatrix} 1 & 2 & 3 & 0 \\ 0 & 2 & -4 & 6 \\ 0 & 0 & -3 & 1 \\ 0 & 0 & 0 & 7 \end{pmatrix} \begin{matrix} \mathbf{r}_1 \\ \mathbf{r}_2 \\ \mathbf{r}_4 \\ \mathbf{r}_3 \end{matrix}, \tag{3.19}$$

which is now in a row-echelon form. The matrix $\mathbf{C}_4$ is a row-echelon form for the original matrix $\mathbf{C}$ in (3.15), as well as for the matrices $\mathbf{C}_1$, $\mathbf{C}_2$, and $\mathbf{C}_3$. All the rows of $\mathbf{C}_4$ are nonzero, hence the rows of the original matrix $\mathbf{C}$ are linearly independent, which means that none of the rows of $\mathbf{C}$ can be written as a linear combination of the other three rows.

We conclude with a few remarks:

1. Although we did not use the elementary row operation of Type 2 explicitly above, we did use it in disguised form. The elementary row operation of Type 3 can be thought of as two separate elementary operations: We first multiply a row by a nonzero scalar (an operation of Type 2), then add the resultant row to another row, which is equivalent to multiplying the resultant row by 1 and adding the result to another row (an operation of Type 3).
2. We can always reduce any matrix to a row-echelon form without ever using any operations of Type 2. Operations of Type 2, however, are needed for other calculations, such as calculating the inverse of square matrices (Chapter 6).
3. As we mentioned earlier, whenever we need to reduce a matrix to a row-echelon form we should work on the matrix systematically so that each time we perform an operation, we do not replace any of the leading zeros by a nonzero number.
4. As stated in Result 3.6, each of the three types of row operations can be carried out either directly, by applying it to the matrix as we have done above, or indirectly, by pre-multiplying the matrix by the corresponding sequence of elementary matrices as illustrated below.

Elementary Matrices

Each elementary row operation corresponds to an elementary matrix. Here are some examples of elementary matrices of order 3×3:

1. An elementary matrix of Type 1,

$$\begin{pmatrix} 0 & 0 & 1 \\ 0 & 1 & 0 \\ 1 & 0 & 0 \end{pmatrix} \begin{matrix} \mathbf{r}_3 \\ \mathbf{r}_2 \\ \mathbf{r}_1 \end{matrix}, \quad \text{obtained by interchanging the first and third rows of } \mathbf{I}_3.$$

2. An elementary matrix of Type 2,

$$\begin{pmatrix} 1 & 0 & 0 \\ 0 & 5 & 0 \\ 0 & 0 & 1 \end{pmatrix} \begin{matrix} \mathbf{r}_3 \\ 5\mathbf{r}_2 \\ \mathbf{r}_1 \end{matrix}, \quad \text{obtained by multiplying the second row of } \mathbf{I}_3 \text{ by 5.}$$

3. An elementary matrix of Type 3,

$$\begin{pmatrix} 1 & 0 & 0 \\ 2 & 1 & 0 \\ 0 & 0 & 1 \end{pmatrix} \begin{matrix} \mathbf{r}_1 \\ \mathbf{r}_2 + 2\mathbf{r}_1 \\ \mathbf{r}_3 \end{matrix}, \text{ obtained by adding to the second row of } \mathbf{I}_3 \text{ twice the first.}$$

To obtain the matrix $\mathbf{C}_1$ in (3.16), we pre-multiply $\mathbf{C}$ in (3.15) by

$$\mathbf{E}_1 = \begin{pmatrix} 0 & 1 & 0 & 0 \\ 1 & 0 & 0 & 0 \\ 0 & 0 & 1 & 0 \\ 0 & 0 & 0 & 1 \end{pmatrix} \begin{matrix} \mathbf{r}_2 \\ \mathbf{r}_1 \\ \mathbf{r}_3 \\ \mathbf{r}_4 \end{matrix},$$

which is an elementary matrix of Type 1 obtained by interchanging the first and second rows of $\mathbf{I}_4$. We obtain

$$\mathbf{E}_1\mathbf{C} = \begin{pmatrix} 0 & 1 & 0 & 0 \\ 1 & 0 & 0 & 0 \\ 0 & 0 & 1 & 0 \\ 0 & 0 & 0 & 1 \end{pmatrix} \begin{pmatrix} 0 & 2 & -4 & 6 \\ 1 & 2 & 3 & 0 \\ 0 & -1 & 2 & 4 \\ -2 & -4 & -9 & 1 \end{pmatrix} = \begin{pmatrix} 1 & 2 & 3 & 0 \\ 0 & 2 & -4 & 6 \\ 0 & -1 & 2 & 4 \\ -2 & -4 & -9 & 1 \end{pmatrix} = \mathbf{C}_1.$$

Note that $\mathbf{E}_1$ is an example of a permutation matrix because its effect on $\mathbf{C}$ is a permutation of its rows; that is, rows 1, 2, 3, and 4 in $\mathbf{C}$ become rows 2, 1, 3, and 4 in $\mathbf{C}_1$. To obtain the matrix $\mathbf{C}_2$ in (3.17), we pre-multiply $\mathbf{C}_1$ by

$$\mathbf{E}_2 = \begin{pmatrix} 1 & 0 & 0 & 0 \\ 0 & 1 & 0 & 0 \\ 0 & 0 & 1 & 0 \\ 0 & 2 & 0 & 1 \end{pmatrix} \begin{matrix} \mathbf{r}_1 \\ \mathbf{r}_2 \\ \mathbf{r}_3 \\ \mathbf{r}_4 + 2\mathbf{r}_1 \end{matrix}$$

which is an elementary matrix of Type 3. We obtain

$$\mathbf{E}_2\mathbf{C}_1 = \begin{pmatrix} 1 & 0 & 0 & 0 \\ 0 & 1 & 0 & 0 \\ 0 & 0 & 1 & 0 \\ 0 & 2 & 0 & 1 \end{pmatrix} \begin{pmatrix} 1 & 2 & 3 & 0 \\ 0 & 2 & -4 & 6 \\ 0 & -1 & 2 & 4 \\ -2 & -4 & -9 & 1 \end{pmatrix} = \begin{pmatrix} 1 & 2 & 3 & 0 \\ 0 & 2 & -4 & 6 \\ 0 & -1 & 2 & 4 \\ 0 & 0 & -3 & 1 \end{pmatrix} = \mathbf{C}_2.$$

To obtain the matrix $\mathbf{C}_3$ in (3.18), we pre-multiply $\mathbf{C}_2$ by

$$\mathbf{E}_3 = \begin{pmatrix} 1 & 0 & 0 & 0 \\ 0 & 1 & 0 & 0 \\ 0 & 0.5 & 1 & 0 \\ 0 & 0 & 0 & 1 \end{pmatrix} \begin{matrix} \mathbf{r}_1 \\ \mathbf{r}_2 \\ \mathbf{r}_3 + 0.5\mathbf{r}_2 \\ \mathbf{r}_4 \end{matrix}$$

and obtain

$$\mathbf{E}_3\mathbf{C}_2 = \begin{pmatrix} 1 & 0 & 0 & 0 \\ 0 & 1 & 0 & 0 \\ 0 & 0.5 & 1 & 0 \\ 0 & 0 & 0 & 1 \end{pmatrix} \begin{pmatrix} 1 & 2 & 3 & 0 \\ 0 & 2 & -4 & 6 \\ 0 & -1 & 2 & 4 \\ 0 & 0 & -3 & 1 \end{pmatrix} = \begin{pmatrix} 1 & 2 & 3 & 0 \\ 0 & 2 & -4 & 6 \\ 0 & 0 & 0 & 7 \\ 0 & 0 & -3 & 1 \end{pmatrix} = \mathbf{C}_3.$$

Finally, we pre-multiply $\mathbf{C}_3$ by

$$\mathbf{E}_4 = \begin{pmatrix} 1 & 0 & 0 & 0 \\ 0 & 1 & 0 & 0 \\ 0 & 0 & 0 & 1 \\ 0 & 0 & 1 & 0 \end{pmatrix} \begin{matrix} \mathbf{r}_1 \\ \mathbf{r}_2 \\ \mathbf{r}_4 \\ \mathbf{r}_3 \end{matrix}$$

and obtain

$$\mathbf{E}_4\mathbf{C}_3 = \begin{pmatrix} 1 & 0 & 0 & 0 \\ 0 & 1 & 0 & 0 \\ 0 & 0 & 0 & 1 \\ 0 & 0 & 1 & 0 \end{pmatrix} \begin{pmatrix} 1 & 2 & 3 & 0 \\ 0 & 2 & -4 & 6 \\ 0 & 0 & 0 & 7 \\ 0 & 0 & -3 & 1 \end{pmatrix} = \begin{pmatrix} 1 & 2 & 3 & 0 \\ 0 & 2 & -4 & 6 \\ 0 & 0 & -3 & 1 \\ 0 & 0 & 0 & 7 \end{pmatrix} = \mathbf{C}_4.$$

The above example illustrates the fact that elementary row operations can be performed by pre-multiplying a matrix by an elementary matrix. We also see that $\mathbf{C}_4$ can be expressed as $\mathbf{E}_4\mathbf{E}_3\mathbf{E}_2\mathbf{E}_1\mathbf{C} = \mathbf{C}_4$, indicating that $\mathbf{C}_4$, which is a row-echelon form of $\mathbf{C}$, can be obtained by pre-multiplying $\mathbf{C}$ by a sequence of elementary matrices as stated in Result 3.6.

We conclude this appendix with two remarks:

1. It is clear that a row-echelon form of a matrix is not unique. For example, multiplying the second row of $\mathbf{C}_4$, which is already in a row-echelon form, by 0.5 (an operation of Type 2) gives

$$\mathbf{C}_5 = \begin{pmatrix} 1 & 2 & 3 & 0 \\ 0 & 1 & -2 & 3 \\ 0 & 0 & -3 & 1 \\ 0 & 0 & 0 & 7 \end{pmatrix},$$

which is also in a row-echelon form. Fortunately, Result 3.1 still holds no matter which row-echelon form we end up with.

2. Equivalent matrices: The matrices $\mathbf{C}$, $\mathbf{C}_1$, $\mathbf{C}_2$, $\mathbf{C}_3$, and $\mathbf{C}_4$ above are not equal, but they are equivalent. Two matrices are said to be *equivalent* if one matrix can be obtained from the other by any sequence of elementary row operations. Thus, a matrix and its row-echelon forms are equivalent. Matrix equivalence should not be confused with matrix equality as defined in Chapter 1.

Exercises

3.1. Express the vector $(5 \quad 2)^T$ as a linear combination of the set of unit vectors.

3.2. In each of the following cases, determine whether the given set of vectors is linearly dependent.

(a) $\mathbf{v}_1 = \begin{pmatrix} 1 \\ 2 \end{pmatrix}, \mathbf{v}_2 = \begin{pmatrix} 3 \\ 6 \end{pmatrix}.$ (b) $\mathbf{v}_1 = \begin{pmatrix} 1 \\ 2 \end{pmatrix}, \mathbf{v}_2 = \begin{pmatrix} 3 \\ -6 \end{pmatrix}.$

(c) $\mathbf{v}_1 = \begin{pmatrix} 1 \\ 2 \end{pmatrix}, \mathbf{v}_2 = \begin{pmatrix} 3 \\ 6 \end{pmatrix}, \mathbf{v}_3 = \begin{pmatrix} 1 \\ 0 \end{pmatrix}.$ (d) $\mathbf{v}_1 = \begin{pmatrix} 1 \\ 2 \end{pmatrix}, \mathbf{v}_2 = \begin{pmatrix} 3 \\ -6 \end{pmatrix}, \mathbf{v}_3 = \begin{pmatrix} 1 \\ 0 \end{pmatrix}.$

(e) $\mathbf{v}_1 = \begin{pmatrix} 2 \\ 1 \\ 2 \end{pmatrix}, \mathbf{v}_2 = \begin{pmatrix} 1 \\ 0 \\ 1 \end{pmatrix}.$ (f) $\mathbf{v}_1 = \begin{pmatrix} 0 \\ 0 \\ 0 \end{pmatrix}, \mathbf{v}_2 = \begin{pmatrix} 1 \\ 0 \\ 1 \end{pmatrix}.$

3.3. For what values of a is each of the following sets linearly dependent?

(a) $\begin{pmatrix} 1 \\ a \end{pmatrix}$ and $\begin{pmatrix} a \\ 1 \end{pmatrix}$ (b) $\begin{pmatrix} 1 \\ a \end{pmatrix}$ and $\begin{pmatrix} 2 \\ a \end{pmatrix}$

3.4. For what values of a and b are the vectors $(1 \quad a \quad 1)^T$ and $(b \quad 1 \quad b)^T$ linearly dependent?

3.5. Determine whether each of the following matrices is in a row-echelon form:

(a) $\mathbf{A} = \begin{pmatrix} 0 & 1 & 0 & 3 & 1 \\ 0 & 0 & 0 & 0 & 0 \\ 0 & 0 & 2 & -1 & 2 \\ 0 & 0 & 0 & 0 & 0 \end{pmatrix}.$ (b) $\mathbf{B} = \begin{pmatrix} 0 & 1 & 0 & 3 & 1 \\ 0 & 0 & 2 & -1 & 2 \\ 0 & 0 & 0 & 0 & 0 \\ 0 & 0 & 0 & 0 & 0 \end{pmatrix}.$

(c) $\mathbf{C} = \begin{pmatrix} 0 & 0 & 0 & 0 \\ 1 & 0 & 0 & 0 \\ 2 & 1 & 0 & 0 \end{pmatrix}.$ (d) $\mathbf{D} = \begin{pmatrix} 0 & 0 & 0 & 1 \\ 0 & 0 & 0 & 0 \\ 0 & 0 & 0 & 0 \end{pmatrix}.$

3.6. Compute the rank of each of the following matrices:

$$\mathbf{A} = \begin{pmatrix} 1 & 2 \\ 2 & 6 \end{pmatrix}, \mathbf{B} = \begin{pmatrix} 1 & 3 \\ 2 & -6 \end{pmatrix}, \mathbf{C} = \begin{pmatrix} 2 & 1 & 2 \\ 1 & 0 & 1 \end{pmatrix}, \mathbf{D} = \begin{pmatrix} 0 & 0 \\ 0 & 0 \\ 0 & 0 \end{pmatrix}, \mathbf{E} = \begin{pmatrix} 0 & 0 & 0 \\ 1 & 0 & 1 \end{pmatrix}.$$

3.7. What are the orders of the matrices **P**, **Q**, and the three null matrices in Result 3.5(h)?

3.8 None of the following matrices is in a row-echelon form, but each requires only one elementary row or column operation to be in a row-echelon form.

$$\mathbf{A} = \begin{pmatrix} 0 & 1 & 0 & 3 & 1 \\ 0 & 0 & 2 & 0 & 0 \\ 0 & 0 & 2 & -1 & 2 \\ 0 & 0 & 0 & 0 & 0 \end{pmatrix}. \qquad \mathbf{B} = \begin{pmatrix} 1 & 1 & 0 & 3 & 0 \\ 2 & 0 & 2 & -1 & 0 \\ 0 & 0 & 0 & 0 & 0 \\ 0 & 0 & 0 & 0 & 0 \end{pmatrix}.$$

$$\mathbf{C} = \begin{pmatrix} 1 & -1 & 1 & 0 \\ 0 & 1 & 1 & 0 \\ 2 & -2 & 0 & 0 \end{pmatrix}. \qquad \mathbf{D} = \begin{pmatrix} 0 & 2 & 0 & 1 \\ 0 & 1 & 0 & 0 \\ 0 & 0 & 0 & 0 \end{pmatrix}.$$

(a) Perform the required row operations that reduce each matrix to a row-echelon form.
(b) What is the corresponding elementary matrix for each case?
(c) Perform the multiplication by the elementary matrix and check that you obtained the desired result.
(d) Which of the matrices has a linearly independent set of row vectors?
(e) How many row vectors are linearly independent in each of the matrices?

3.9. Show that any set of vectors that includes one or more null vectors must be linearly dependent.

3.10. What are the conditions under which each of the following matrices is in a row-echelon form?
(a) An identity matrix
(b) A diagonal matrix
(c) A null matrix
(d) A matrix of ones
(e) A symmetric matrix
(f) A skew-symmetric matrix
(g) A lower-triangular matrix
(h) An upper-triangular matrix
(i) An elementary matrix
(j) A permutation matrix

3.11. In each of the following cases, find the permutation matrices required to transform a 3×4 matrix **A** to a matrix **B**, where
(a) The rows of **B** are the same as those of **A** but given in reverse order.
(b) The rows of **B** are the same as those of **A** but given in the order 3, 1, 4, 2.
(c) The columns of **B** are the same as those of **A** but given in the order 1, 3, 2, 4.
(d) **B** is obtained from **A** by first reversing its rows and then reversing the columns of the resultant matrix.
(e) **B** = **A**.

3.12. Given the following matrices:

$$\mathbf{A} = \begin{pmatrix} 1 & 2 & 0 & 1 \\ 0 & 2 & 1 & 5 \\ 4 & 0 & -1 & 0 \end{pmatrix}, \qquad \mathbf{B} = \begin{pmatrix} 0 & 0 & 0 & 2 \\ 1 & 2 & 0 & 1 \\ 2 & 0 & 1 & 2 \end{pmatrix},$$

$$\mathbf{C} = \begin{pmatrix} 1 & -1 & 1 & 0 \\ 0 & 1 & 1 & 0 \\ 2 & -2 & 0 & 0 \end{pmatrix}, \qquad \mathbf{D} = \begin{pmatrix} 0 & 2 & 5 & 2 \\ 1 & 0 & 3 & 1 \\ -1 & 2 & 2 & 1 \end{pmatrix};$$

(a) Find a row-echelon form for each matrix.
(b) Which matrices contain a linearly independent set of row vectors?
(c) Which matrices contain a linearly independent set of column vectors? [Hint: Use Result 3.1(e).]

3.13. A matrix **A**, consisting of five rows and nine columns, contains no null row or column vectors. One of its row-echelon forms contains no null rows. State whether each of the following statements is true or false (explain).

(a) **A** contains a linearly independent set of row vectors.
(b) **A** contains a linearly independent set of column vectors.
(c) All row-echelon forms of **A** contain no null row vectors.
(d) The number of linearly independent rows of **A** is five.
(e) The number of linearly independent columns of **A** is nine.
(f) Every column-echelon form of **A** contains no null vectors.
(g) One row of **A** can be expressed as a linear combination of the other rows.
(h) One column of **A** can be expressed as a linear combination of the other columns.
(i) An elementary matrix is a matrix obtained from **A** after performing an elementary row operation on its rows.
(j) A permutation matrix is a matrix obtained from **A** after permuting its rows and/or its columns.
(k) Any elementary matrix required to permute the rows of **A** has to be of order 5×5.
(l) Any elementary matrix required to permute the columns of **A** has to be of order 9×5.
(m) **A** contains at least five linearly independent columns.
(n) A permutation of the rows of **A** can be obtained by pre-multiplying **A** by an appropriate permutation matrix.

3.14. A matrix **A**, consisting of five rows and nine columns, contains no null row or column vectors. One of its row-echelon forms contains one null row and one null column. State whether each of the statements in Exercise 3.13 is true or false (explain).

4 Vector Geometry

4.1. Introduction

Geometry and matrix algebra are intimately related and matrix concepts can often be easily understood by relating them to their geometric counterparts. This chapter discusses some geometric and graphic aspects of vectors. To be specific, in Sections 4.2 and 4.3 we describe three different but equivalent ways of viewing vectors: algebraically, graphically, and geometrically. In Section 4.4 we show the connection between the geometric representation of vectors and the concepts of linear dependence and independence discussed in Chapter 3. In Sections 4.5 and 4.6 we discuss the geometry of vector algebra and the notion of a vector space.

4.2. Algebraic and Graphic Views of Vectors

We have already defined a vector algebraically as a matrix with one row or one column. We can also think of a vector, graphically, as a point in a space, where the elements of the vector are the coordinates of the point. Consider, for example, the following vectors:

$$\mathbf{a} = \begin{pmatrix} 5 \\ 0 \end{pmatrix}, \quad \mathbf{b} = \begin{pmatrix} 4 \\ 4 \end{pmatrix}, \quad \mathbf{c} = \begin{pmatrix} -2 \\ 2 \end{pmatrix}, \quad \mathbf{d} = \begin{pmatrix} -4 \\ 0 \end{pmatrix}. \tag{4.1}$$

Each of these vectors has two elements. If we regard, for example, the first element, $a_1 = 5$, as the x-coordinate and the second element, $a_2 = 0$, as the y-coordinate, each vector can be represented by a point in a two-dimensional space as shown in Figure 4.1. This idea can be extended to vectors with three or more elements, but we need a space of three or more dimensions to do so. Our inability to visualize a multidimensional space, however, should not prevent us from thinking of any n-element vector as a point in an n-dimensional space.

4.3. Geometric Views of Vectors

A vector can also be viewed geometrically as a directed line segment with an arrow starting from the origin and ending at the point representing the vector in the space. This geometric representation of the vectors in (4.1) is shown in Figure 4.2.

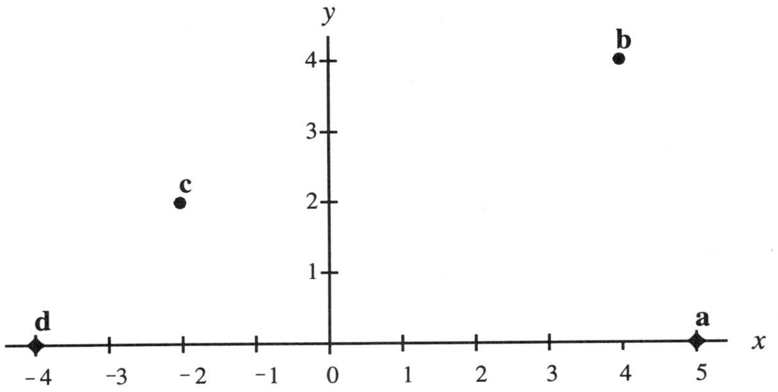

Figure 4.1. A graphic view of vectors. The vectors **a**, **b**, **c**, and **d** in (4.1) are represented as points in a two-dimensional space.

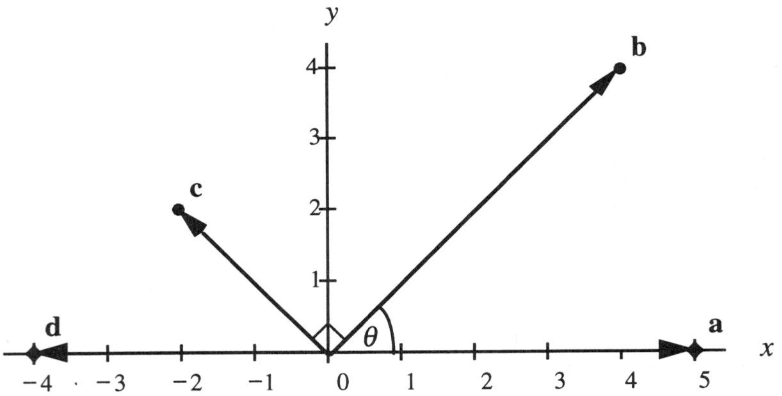

Figure 4.2. A geometric view of vectors. The vectors **a**, **b**, **c**, and **d** in (4.1) are represented as line segments, starting at the origin and ending at the point representing the vector in the space.

Again, this way of viewing vectors can be extended to vectors with three or more elements, as we can always imagine a line segment starting at the origin of an n-dimensional space and ending at the point representing the vector in the space.

This geometric representation of vectors raises two natural questions:
(a) What are the length and direction of the line segment representing a vector?
(b) What is the angle between two line segments representing two vectors?
These questions are addressed next.

4.3.1. Length or Magnitude of a Vector

We now have three ways of viewing a vector: algebraically, graphically, and geometrically. These seemingly different views of vectors are of course related. For example, let us measure the length of the line segment representing the vector **a**. Using the Pythagorean theorem, the length of the vector **a** (that is, the line from the origin to the point **a** in Figure 4.2) is

$$\sqrt{5^2 + 0^2} = +\sqrt{25} = 5.$$

Since the length is a nonnegative quantity, only the positive square-root is considered here. Similarly, the length of the vector **b** is

$$\sqrt{4^2 + 4^2} = \sqrt{32} = 4\sqrt{2}\,.$$

The reader can verify that the lengths of the vectors **c** and **d** are $2\sqrt{2}$ and 4, respectively.

More generally, for any n-dimensional vector

$$\mathbf{v} = \begin{pmatrix} v_1 \\ v_2 \\ \vdots \\ v_n \end{pmatrix},$$

the *length* of **v**, denoted by $\| \mathbf{v} \|$, is defined as

$$\| \mathbf{v} \| = \sqrt{\mathbf{v}^{\mathrm{T}}\mathbf{v}} = \sqrt{v_1^2 + v_2^2 + \cdots + v_n^2} = \left(\sum_{i=1}^{n} v_i^2 \right)^{1/2}. \tag{4.2}$$

You may think of (4.2) as the generalization of the Pythagorean theorem to the n-dimensional space. Although we may not be able to draw the line segment representing a vector in an n-dimensional space, we can always measure its length using (4.2). For example, the lengths of the vectors

$$\mathbf{e} = \begin{pmatrix} 2 \\ -1 \\ 0 \\ -2 \end{pmatrix} \quad \text{and} \quad \mathbf{f} = \begin{pmatrix} 0.5 \\ 0.5 \\ -0.5 \\ 0.5 \end{pmatrix} \tag{4.3}$$

are

$$\| \mathbf{e} \| = \sqrt{\mathbf{e}^T\mathbf{e}} = \sqrt{2^2 + (-1)^2 + 0^2 + (-2)^2} = \sqrt{9} = 3$$

and

$$\| \mathbf{f} \| = \sqrt{\mathbf{f}^T\mathbf{f}} = \sqrt{0.5^2 + 0.5^2 + (-0.5)^2 + 0.5^2} = 1,$$

respectively. The length of a vector is also referred to as the *norm* of the vector.[1] Note that the length of a geometric vector can be interpreted as the *magnitude* of the corresponding algebraic vector. Thus, the larger the elements of the vector in absolute value, the greater its length and the larger the magnitude of the vector. Therefore, the words *length*, *norm*, and *magnitude* can be used synonymously because they measure the same characteristic of a vector.

4.3.2. Angle Between Two Vectors

To further illustrate the relationships among the three ways of viewing vectors discussed earlier, let us measure the angle between any two geometric vectors. Refer first to the triangle in Figure 4.3 and recall that the cosine of the angle, θ, between the vectors $\mathbf{u}$ and $\mathbf{v}$ is defined, in terms of the x and y coordinate values, as $\cos \theta = x/r$. The angle between two vectors can be measured using the next result.

Result 4.1. Let θ be the angle between two vectors $\mathbf{u}$ and $\mathbf{v}$. Then

$$\cos \theta = \frac{\mathbf{u}^T\mathbf{v}}{\|\mathbf{u}\| \times \|\mathbf{v}\|}.$$

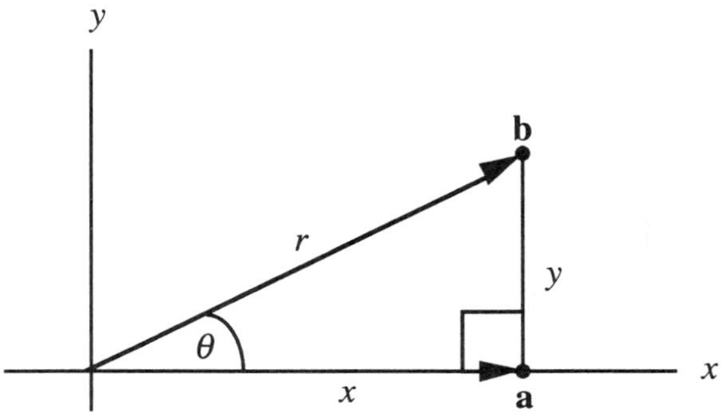

Figure 4.3. The cosine of the angle θ is defined as $\cos \theta = x/r$.

---

1 Actually, the words *length* and *norm* are used interchangeably. The concept of vector norm is discussed in more detail in Section 5.3.

The numerator is the inner product of the vectors **u** and **v**. Thus, the cosine of the angle between two vectors is the ratio of their inner product and the product of their lengths (norms). For example, let us use Result 4.1 to measure the angle between the vectors **a** and **b** in Figure 4.2. We have

$$\mathbf{a}^{\mathrm{T}}\mathbf{b} = 5 \times 4 + 0 \times 4 = 20,$$

$$\| \mathbf{a} \| = \sqrt{5^2 + 0^2} = 5,$$

$$\| \mathbf{b} \| = \sqrt{4^2 + 4^2} = 4\sqrt{2},$$

and hence

$$\cos \theta = \frac{\mathbf{a}^{\mathrm{T}}\mathbf{b}}{\|\mathbf{a}\| \times \|\mathbf{b}\|} = \frac{20}{5 \times 4\sqrt{2}} = \frac{1}{\sqrt{2}}.$$

Now, since $\cos \theta = 1/\sqrt{2}$ and θ is an acute angle, it follows that $\theta = 45°$, as would be expected from Figure 4.2. As a reminder, Table 4.1 gives the cosines of some selected angles.

Table 4.1. Cosines of Certain Angles

θ	$0°$	$30°$	$45°$	$60°$	$90°$	$120°$	$135°$	$150°$	$180°$
$\cos \theta$	1	$\frac{\sqrt{3}}{2}$	$\frac{1}{\sqrt{2}}$	$\frac{1}{2}$	0	$-\frac{\sqrt{3}}{2}$	$-\frac{1}{\sqrt{2}}$	$-\frac{1}{2}$	-1

From the above discussion, we can see that every vector has two attributes: a *magnitude* measured by its length and a *direction* measured by the angles it makes with the principal axes (the x- and y-axes in two dimensions). Thus, for example, the vector **a** in Figure 4.2 has a magnitude of 5 and makes an angle of $0°$ with the (positive side of the) x-axis and an angle of $90°$ with the (positive side of the) y-axis, whereas the vector **b** in Figure 4.2 has a magnitude of $\sqrt{32} = 4\sqrt{2}$ and makes an angle of $45°$ with the x-axis and an angle of $45°$ with the y-axis.

4.4. Vector Geometry and Linear Dependence

The angle between two vectors is directly related to the concepts of linear dependence and independence discussed in Chapter 3. Recall from Section 3.2 that two vectors **u** and **v** are linearly dependent if and only if one vector can be written as a nonzero scalar multiple of the other. Let us now use Result 4.1 to measure the angle θ between two dependent vectors $\mathbf{u}$ and $\mathbf{v} = a\mathbf{u}$. Note that a is a scalar, so we have, $\mathbf{u}^T\mathbf{v} = a\mathbf{u}^T\mathbf{u} = a \parallel \mathbf{u} \parallel^2$ and $\mathbf{v}^T\mathbf{v} = a^2\mathbf{u}^T\mathbf{u}$, hence $\parallel \mathbf{v} \parallel = |a| \times \parallel \mathbf{u} \parallel$, where $|a|$ means the absolute value of a. Therefore,

$$\cos\theta = \frac{\mathbf{u}^T\mathbf{v}}{\|\mathbf{u}\| \times \|\mathbf{v}\|} = \frac{a\|\mathbf{u}\|^2}{|a| \times \|\mathbf{u}\| \times \|\mathbf{u}\|} = \pm 1,$$

which implies that $\theta = 0^o$ or 180° (see Table 4.1). We may then conclude that when two vectors are linearly dependent, the angle between them is either 0° or 180°, which means that the two line segments representing the vectors have the same (or opposite) direction. The converse of this statement is also true. This is stated in the following result.

Result 4.2. Two vectors of the same order are linearly dependent if and only if the angle between them is either 0° or 180°; otherwise the vectors are linearly independent.[2]

For example, let us use Result 4.1 to measure the angle between the vectors **a** and **d** in Figure 4.2, which can be seen to be 180°. We have

$$\cos\theta = \frac{\mathbf{a}^T\mathbf{d}}{\|\mathbf{a}\| \times \|\mathbf{d}\|} = \frac{-20}{5 \times 4} = -1,$$

which implies that $\theta = 180^o$ as can be seen from Figure 4.2. Thus, by Result 4.2, the vectors **a** and **d** are linearly dependent. This can also be seen from the fact that $\mathbf{a} = -1.25\ \mathbf{d}$; that is, one vector can be written as a nonzero scalar multiple of the other.

Let us now measure the angle between the vectors **b** and **c**. Since $\mathbf{b}^T\mathbf{c} = 4 \times (-2) + (4 \times 2) = 0$, we have

$$\cos\theta = \frac{\mathbf{b}^T\mathbf{c}}{\|\mathbf{b}\| \times \|\mathbf{c}\|} = \frac{0}{4\sqrt{2} \times 2\sqrt{2}} = 0. \qquad (4.4)$$

Thus, the angle between **b** and **c** is 90° as can be seen from Figure 4.2. By Result 4.2, the vectors **b** and **c** are linearly independent (neither of the two vectors can be written as a scalar multiple of the other). Furthermore, from (4.4), we can deduce the following result.

Result 4.3. Two vectors **u** and **v** of the same order are *perpendicular* if and only if $\mathbf{u}^T\mathbf{v} = 0$.

---

2 For simplicity and clarity of exposition, in this book we shall not be concerned with the equivalent angles obtained by adding multiples of 360°. Thus, for example, when we say an angle is 90°, we also mean $90^o + 360^o$, $90^o + 720^o$, etc.

Recall from Chapter 2 that when $\mathbf{u}^T\mathbf{v} = 0$, $\mathbf{u}$ and $\mathbf{v}$ are said to be orthogonal. Thus, we can conclude that if two algebraic vectors are orthogonal, then the two line segments representing them are perpendicular. The converse is also true, and hence the words *orthogonal* and *perpendicular* are used interchangeably. We therefore have the following result.

Result 4.4. Let $\mathbf{u}$ and $\mathbf{v}$ be two vectors of the same order; then
(a) $\mathbf{u}$ and $\mathbf{v}$ are orthogonal if and only if the angle between them is 90°.
(b) if $\mathbf{u}$ and $\mathbf{v}$ are orthogonal, then they are linearly independent.

Note, however, that if two vectors are linearly independent they may or may not be orthogonal. To summarize, when speaking of the angle θ between two vectors, we have three possibilities:

1. $\theta = 0°$ or 180°. In this case the vectors have the same or opposite direction and one can be expressed as a nonzero scalar multiple of the other, which means that they are linearly dependent .
2. $\theta = 90°$. In this case the vectors are orthogonal, hence linearly independent.
3. θ is different from 0°, 90°, and 180°. In this case the vectors are linearly independent but not orthogonal.

4.5. Geometry of Vector Algebra

Let us now take a look at what happens graphically when we add or subtract two vectors, when we multiply a vector by a scalar (in particular, when we normalize a vector), or when we multiply a vector by a matrix.

4.5.1. Geometry of Vector Addition

Suppose we wish to add the vectors

$$\mathbf{b} = \begin{pmatrix} 4 \\ 4 \end{pmatrix} \quad \text{and} \quad \mathbf{c} = \begin{pmatrix} -2 \\ 2 \end{pmatrix}.$$

Algebraically, we have

$$\mathbf{e} = \mathbf{b} + \mathbf{c} = \begin{pmatrix} 2 \\ 6 \end{pmatrix}.$$

Geometrically, the vectors $\mathbf{b}$, $\mathbf{c}$, and $\mathbf{e}$ are drawn in Figure 4.4. We see that the line segment representing the vector $\mathbf{e}$, which is the sum of $\mathbf{b}$ and $\mathbf{c}$, is the diagonal of the parallelogram having the line segments representing $\mathbf{b}$ and $\mathbf{c}$ as adjacent edges. This is known as the *parallelogram rule*, and is always true for the sum of any two vectors.

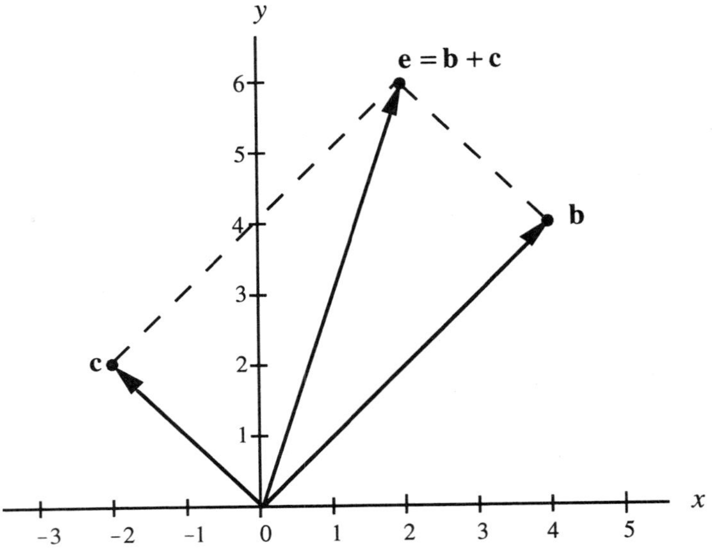

Figure 4.4. The effect of adding two vectors. The sum of the vectors **b** and **c** is a vector **e** which is the diagonal of the parallelogram having the line segments representing **b** and **c** as adjacent edges.

4.5.2. Geometry of Vector Subtraction

What about the difference between two vectors? The vector

$$\mathbf{g} = \mathbf{b} - \mathbf{c} = \begin{pmatrix} 6 \\ 2 \end{pmatrix} \tag{4.5}$$

is shown in Figure 4.5. The length of **g** is $\|\mathbf{g}\| = \sqrt{40}$. This is also equal to the dashed line connecting the points **b** and **c** in Figure 4.5. Thus, the distance between the vectors **b** and **c** is the same as the norm (length) of the vector **g**, which represents the difference between the two vectors. The notion of distance plays a major role in many applications, such as multivariate and regression analyses, and in statistics in general. For this reason, Section 10.2 is devoted to a discussion of the concept of distance.

It is sometimes useful to consider the difference **b** – **c** as the sum of **b** and –**c**. Figure 4.6 is the same as Figure 4.5 but with the vector –**c** added. As would be expected, we see that the vector **g** is the diagonal of the parallelogram having the line segments representing **b** and –**c** as adjacent edges.

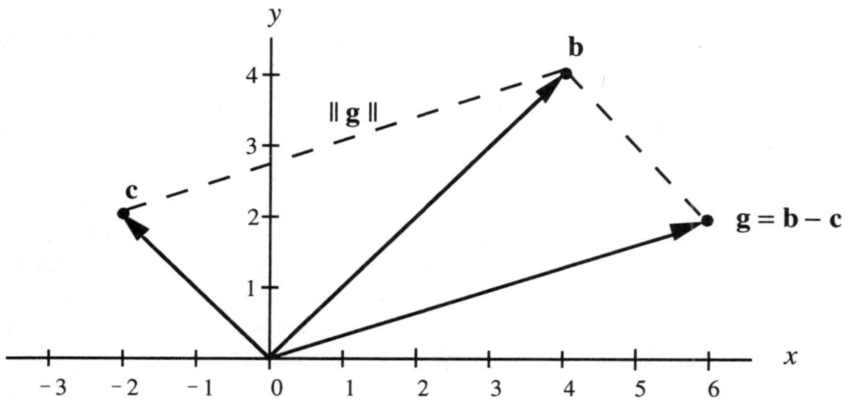

Figure 4.5. The effect of subtracting one vector from another. The norm of the vector **g** = **b** − **c** is equal to the distance between the two points representing the vectors **b** and **c** (the dashed line between **b** and **c**).

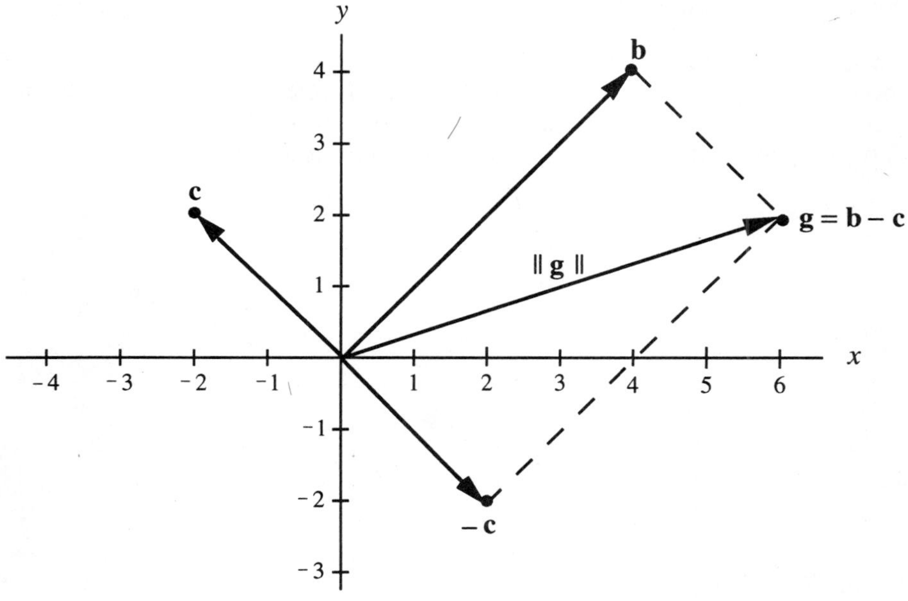

Figure 4.6. The difference between two vectors **b** and **c** can be viewed as the sum of **b** and −**c**. Thus, the sum of the two vectors **b** and −**c** is **g** = **b** + (−**c**). The vector **g** is the diagonal of the parallelogram having the line segments representing **b** and −**c** as adjacent edges.

4.5.3. Geometry of Multiplying a Vector by a Scalar

Multiplying a vector by a scalar amounts to multiplying each of its elements by the scalar. Consider, for example, the four vectors

$$\mathbf{a} = \begin{pmatrix} 4 \\ 2 \end{pmatrix}, \quad \mathbf{b} = 2\mathbf{a} = \begin{pmatrix} 8 \\ 4 \end{pmatrix}, \quad \mathbf{c} = 0.5\mathbf{a} = \begin{pmatrix} 2 \\ 1 \end{pmatrix}, \quad \mathbf{d} = -0.5\mathbf{a} = \begin{pmatrix} -2 \\ -1 \end{pmatrix}.$$

The vectors **b**, **c**, and **d** are obtained by multiplying **a** by nonzero scalars. The geometric effects of these multiplications are shown in Figure 4.7. We see, for example, that the effect of multiplying **a** by 2 is simply doubling its length. The length of **a** is $\|\mathbf{a}\| = \sqrt{20}$ and the length of **b** is $\|\mathbf{b}\| = \sqrt{80} = 2\sqrt{20}$. In general, $\| s\mathbf{v} \| = |s| \times \|\mathbf{v}\|$, for any scalar s. In Figure 4.7, we can see that the length of **b** is indeed twice the length of **a** (because $|s| = 2$), and that the lengths of **c** and **d** are half that of **a** (because in both cases $|s| = 0.5$). Note also that **a** , **b**, and **c** have the same direction and **d** has the opposite direction. Therefore, the four vectors are linearly dependent. In fact, any pair of vectors in Figure 4.7 is linearly dependent.

In general, the effect of multiplying a vector by a positive number s is to either stretch (when $s > 1$) or shrink (when $s < 1$) the length of the vector. The resultant vector has the same direction as the original one. If s is negative, the new vector has an opposite direction and its length is multiplied by $|s|$.

Normalizing Vectors. A vector of length (norm) 1 is called a *normal* vector. Any nonzero vector can be *normalized* (i.e., transformed into a normal vector) by dividing each of its elements by its norm. This transformation is called *normalization.* Two properties of vector normalization are stated next.

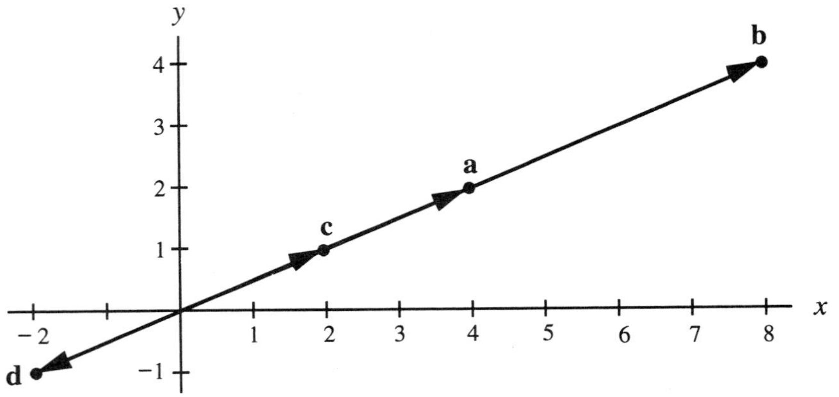

Figure 4.7. Multiplying a vector by a positive scalar shrinks or expands the vector in the same direction, whereas multiplying a vector by a negative scalar shrinks or expands the vector in the opposite direction. For example, the vector **b** = 2**a** has the same direction as **a** and twice its length.

Result 4.5.
(a) The vector $\mathbf{z} = \mathbf{v} / \| \mathbf{v} \| = \| \mathbf{v} \|^{-1} \mathbf{v}$ is normal for any nonzero vector $\mathbf{v}$.
(b) The normalized versions of two vectors of the same direction are equal.

Result 4.5(a) states that any nonzero vector can be normalized simply by dividing each of its elements by its length. For example, the vector

$$\mathbf{e} = \begin{pmatrix} 2 \\ -1 \\ 0 \\ -2 \end{pmatrix}$$

can be normalized by dividing each of its elements by its norm. Since $\| \mathbf{e} \| = 3$, its normalized version is

$$\mathbf{z} = \| \mathbf{e} \|^{-1} \mathbf{e} = \frac{1}{3} \begin{pmatrix} 2 \\ -1 \\ 0 \\ -2 \end{pmatrix} = \begin{pmatrix} 2/3 \\ -1/3 \\ 0/3 \\ -2/3 \end{pmatrix}$$

You may verify that $\| \mathbf{z} \| = 1$.

To illustrate Result 4.5(b), we use the vectors

$$\mathbf{a} = \begin{pmatrix} 4 \\ 3 \end{pmatrix} \quad \text{and} \quad \mathbf{b} = \begin{pmatrix} 8 \\ 6 \end{pmatrix},$$

which have the same direction (why?). Their norms are 5 and 10, and their normalized versions are

$$\| \mathbf{a} \|^{-1} \mathbf{a} = \frac{1}{5} \begin{pmatrix} 4 \\ 3 \end{pmatrix} = \begin{pmatrix} 0.8 \\ 0.6 \end{pmatrix} \quad \text{and} \quad \| \mathbf{b} \|^{-1} \mathbf{b} = \frac{1}{10} \begin{pmatrix} 8 \\ 6 \end{pmatrix} = \begin{pmatrix} 0.8 \\ 0.6 \end{pmatrix},$$

respectively. Thus, the normalized versions of **a** and **b** are identical.

The elements of a normal vector can be interpreted as *direction cosines*. Simply stated, they are the cosines of the angles between the vector and each of the coordinate axes. This is perhaps best illustrated by an example. The four vectors

$$\mathbf{a} = \begin{pmatrix} 3 \\ \sqrt{3} \end{pmatrix}, \quad \mathbf{z} = \begin{pmatrix} \sqrt{3}/2 \\ 1/2 \end{pmatrix}, \quad \mathbf{h} = \begin{pmatrix} 3 \\ 0 \end{pmatrix}, \quad \mathbf{v} = \begin{pmatrix} 0 \\ 2 \end{pmatrix},$$

are drawn in Figure 4.8. The vectors **h** and **v** coincide with the horizontal and vertical axes, respectively. The vector **z** is the normalized version of **a**. Since **a** and **z** are proportional, they have the same direction in Figure 4.8.

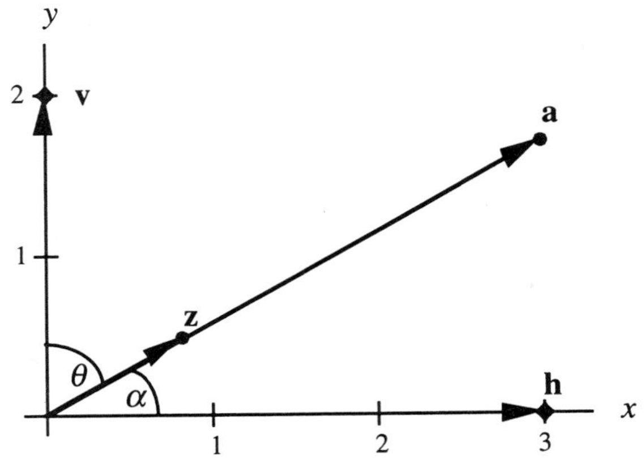

Figure 4.8. The elements of a normalized vector **z** are the direction cosines of the angles α and θ between **z** and the coordinate axes.

Now let us compute the angle α between **z** and **h** (the horizontal axis) and the angle θ between **z** and **v** (the vertical axis). By making use of Result 4.1,

$$\cos\alpha = \frac{\mathbf{z}^{\mathrm{T}}\mathbf{h}}{\|\mathbf{z}\| \times \|\mathbf{h}\|} = \frac{3\sqrt{3}/2}{(1)(3)} = \frac{\sqrt{3}}{2},$$

and

$$\cos\theta = \frac{\mathbf{z}^{\mathrm{T}}\mathbf{v}}{\|\mathbf{z}\| \times \|\mathbf{v}\|} = \frac{1}{(1)(2)} = \frac{1}{2},$$

which implies that $\alpha = 30^{\circ}$ and $\theta = 60^{\circ}$. Note that $\alpha + \theta = 90^{\circ}$, as expected.

The important thing to notice here is that the elements of the normalized vector **z** are nothing but the direction cosines, that is,

$$\mathbf{z} = \begin{pmatrix} \sqrt{3}/2 \\ 1/2 \end{pmatrix} = \begin{pmatrix} \cos\alpha \\ \cos\theta \end{pmatrix}.$$

This is true for normal vectors of any dimensions, as stated in the next result.

Result 4.6. Let **v** be any n-dimensional vector and **z** be the normalized version of **v**, then

$$\mathbf{z} = \begin{pmatrix} \cos\theta_1 \\ \cos\theta_2 \\ \vdots \\ \cos\theta_n \end{pmatrix},$$

where θ_j is the angle between $\mathbf{v}$ and the jth coordinate axis.

The vector norm or length has many other properties, some of which are presented in Section 5.3. Practical examples illustrating the usefulness of the above results are given in Chapters 9 and 11.

4.5.4. Geometry of Multiplying a Vector by a Matrix

We have seen that, from the algebraic point of view, multiplying a vector by a scalar amounts to multiplying each of its elements by the scalar. Alternatively, from the geometric point of view, multiplying a vector by a positive scalar shrinks or expands the vector in the same direction, whereas multiplying a vector by a negative scalar shrinks or expands the vector in the opposite direction. We now discuss the effects of multiplying a vector by a matrix.

Suppose that $\mathbf{A}$ is a matrix and $\mathbf{x}$ is a vector such that the product $\mathbf{y} = \mathbf{A}\mathbf{x}$ exists; then $\mathbf{y}$ is a vector with the same number of elements as the number of rows in $\mathbf{A}$. Thus, $\mathbf{A}$ transforms the vector $\mathbf{x}$ into another vector $\mathbf{y}$. This type of transformation is called a *linear transformation*. We shall study linear transformations in more detail in Chapter 7, but now we are interested in the geometric effects of multiplying a vector by a matrix. For example, let $\mathbf{y} = \mathbf{A}\mathbf{x}$, where

$$\mathbf{A} = \begin{pmatrix} a & b \\ c & d \end{pmatrix} \quad \text{and} \quad \mathbf{x} = \begin{pmatrix} x_1 \\ x_2 \end{pmatrix}.$$

Then,

$$\mathbf{y} = \begin{pmatrix} y_1 \\ y_2 \end{pmatrix} = \begin{pmatrix} a\,x_1 + b\,x_2 \\ c\,x_1 + d\,x_2 \end{pmatrix},$$

from which we can see that $y_1 = a\,x_1 + b\,x_2$ and $y_2 = c\,x_1 + d\,x_2$. Hence, $\mathbf{y}$ is a linear transformation of $\mathbf{x}$. As a numerical example, consider

$$\mathbf{A} = \begin{pmatrix} 1 & 1 \\ 1 & -1 \end{pmatrix} \quad \text{and} \quad \mathbf{x} = \begin{pmatrix} 3 \\ 2 \end{pmatrix}.$$

Then,

$$\mathbf{y} = \mathbf{A}\mathbf{x} = \begin{pmatrix} 1 & 1 \\ 1 & -1 \end{pmatrix}\begin{pmatrix} 3 \\ 2 \end{pmatrix} = \begin{pmatrix} 5 \\ 1 \end{pmatrix}.$$

The vectors **x** and **y** are drawn in Figure 4.9, from which we see that the effect of multiplying **x** by **A** is to obtain another vector **y** different from **x** in both magnitude and direction. Thus, you may think of **A** as a *rotation* matrix because when multiplied by **A**, the vector **x** is rotated to **y**. (More is said about rotation matrices in Chapters 7 and 11.) Thus, we conclude that multiplying a vector by a scalar can change the magnitude of the vector, but the direction of the new vector can only be either the same or directly opposite the direction of the original vector; in contrast, multiplying a vector by a matrix can change both its magnitude and direction.

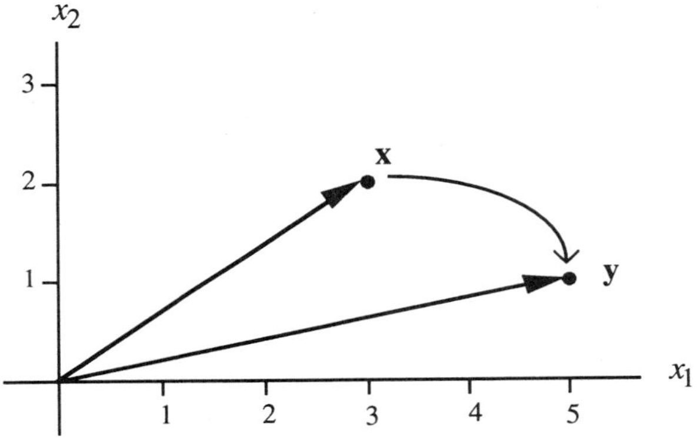

Figure 4.9. The vector **y** is obtained by multiplying the vector **x** by the matrix **A**. The magnitude and direction of **y** are different from those of **x**.

4.6. Vector Space

The concept of a vector space is not easy to explain or to grasp. To keep the discussion at a level consistent with the objective of the book, our discussion of the topic has to be selective and the terminology has to be simple. Readers who are interested in a more comprehensive coverage are referred to the books mentioned in Chapter 1.

As we have seen in Figures 4.1 and 4.2, any vector of order 2 can be represented by a point in a two-dimensional space (the Cartesian space). The elements of the vector are the coordinates of the point in the space. These coordinates are measured with respect to an *origin* (the point (0, 0)) and a set of two perpendicular (orthogonal) axes (the x-axis and y-axis). Similarly, any vector of order 3 can be represented by a point in a three-dimensional space. The elements of the vector are the coordinates of the point in the space. These coordinates are measured with respect to an *origin* (the point (0, 0, 0)) and a set of three perpendicular axes (the x-axis, y-axis, and z-axis).

As mentioned in Section 4.2, this idea can be extended to vectors of order n. Thus, any vector of order n can be represented by a point in an n-dimensional space. This space is known as the n-dimensional *Euclidean space* and is denoted by $\Re^n$ (pronounced as $\Re$-n, not as $\Re$ to the power n). Again, the elements of the vector are the coordinates of the point. These coordinates are measured with respect to an *origin* (the null vector of order n) and a set of n perpendicular axes.

Recall from Result 3.3 that any vector of order n can be expressed as a linear combination of n unit vectors of order n. The reason for this is that the set of unit vectors is linearly independent (because the vectors are orthogonal). In fact, any set of n linearly independent vectors (not necessarily the set of unit vectors) can serve the same purpose. This is stated in the following result.

Result 4.7. Any vector $\mathbf{v}$ of order n can be expressed as a linear combination of any set of n linearly independent vectors; that is,

$$\mathbf{v} = a_1\, \mathbf{x}_1 + a_2\, \mathbf{x}_2 + \cdots + a_n\, \mathbf{x}_n,$$

where a_1 , a_2, ..., a_n are scalars and $\mathbf{x}_1$, $\mathbf{x}_2$, ..., $\mathbf{x}_n$ are any linearly independent vectors.

The implication of Result 4.7 is that any vector in $\Re^n$ can be written as a linear combination of any set of n linearly independent vectors. For this reason, such a set of n linearly independent vectors is said to *span* or *generate* $\Re^n$ and is referred to as a *spanning set* of $\Re^n$. For example, the vectors

$$\mathbf{a} = \begin{pmatrix} 1 \\ 3 \end{pmatrix}, \quad \mathbf{b} = \begin{pmatrix} 4 \\ 2 \end{pmatrix}, \quad \text{and} \quad \mathbf{c} = \begin{pmatrix} 3 \\ 3 \end{pmatrix}$$

are pairwise linearly independent. (This can be seen from Figure 4.10, which shows that the three vectors have different directions.) Therefore, any two of the three vectors can generate $\Re^2$ and the third can be written as a linear combination of the other two. For example, the reader can check that $\mathbf{c}$ can be written as $\mathbf{c} = 0.6\,\mathbf{a} + 0.6\,\mathbf{b}$. In fact, *any* two-dimensional vector can be written as a linear combination of $\mathbf{a}$ and $\mathbf{b}$. On the other hand, by graphing the vectors

$$\mathbf{d} = \begin{pmatrix} 2 \\ 1 \end{pmatrix} \quad \text{and} \quad \mathbf{e} = \begin{pmatrix} 4 \\ 2 \end{pmatrix},$$

the reader can see that they have the same direction, hence they are linearly dependent. Therefore, $\mathbf{d}$ and $\mathbf{e}$ span $\Re^1$, not $\Re^2$.

Note that if $m \geq n$, any set of m vectors also spans $\Re^n$ as long as it contains a subset of n linearly independent vectors. The reason is that the set of n linearly independent vectors spans $\Re^n$ and the remaining $m - n$ vectors can be expressed as linear combinations of the n linearly independent vectors. Thus, a spanning set is not unique. When $m = n$, a spanning set is also called a *basis* for $\Re^n$. Thus, a basis set is also a spanning set (although the converse is not necessarily true), and any set of n

linearly independent vectors can serve as a basis set for $\Re^n$. Thus, a basis set is not unique. For example, since the unit vectors $\mathbf{u}_1, \mathbf{u}_2, \ldots, \mathbf{u}_n$ are linearly independent, they form a basis for $\Re^n$; that is, any vector in $\Re^n$ can also be written as a linear combination of the unit vectors. The set of n unit vectors is called the *standard basis* for $\Re^n$.

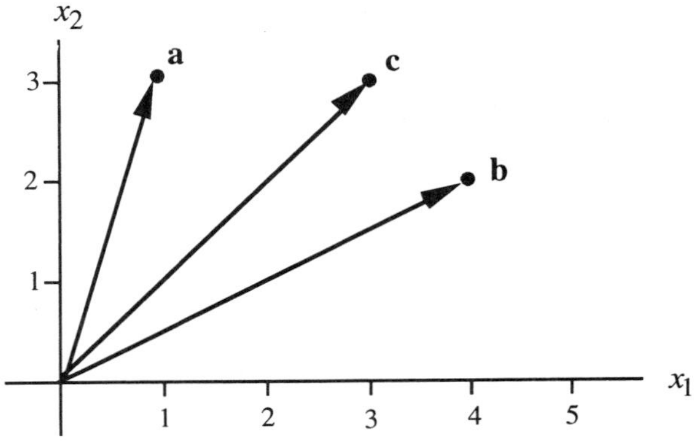

Figure 4.10. The vectors **a** and **b** are linearly independent, hence they span a space of dimension 2. The vector **c**, or any other two-dimensional vector, can be written as a linear combination of **a** and **b**.

Other implications of Result 4.7 are the following:

1. Any nonzero vector generates $\Re^1$ (or $\Re$, for simplicity). This is a space of dimension 1. It consists of all points on a straight line passing through the origin, that is, the set of all real numbers.
2. Any two linearly independent vectors generate $\Re^2$, a space of dimension 2. It is a plane passing through the origin. ($\Re^2$ is the set of all ordered pairs of real numbers.)
3. Any three linearly independent vectors generate $\Re^3$, which is a three-dimensional (physical) space. ($\Re^3$ is the set of all ordered triples of real numbers.)

And so on. A space that is generated or spanned by a set of vectors is called a *vector space*. Thus, for example, in a set of n vectors, each of order n, if the maximum number of linearly independent vectors is k, then the set generates a space of dimension k. This fact is stated in the following result.

Result 4.8. The space spanned by the columns of a matrix **A**, referred to as the *column space* of **A**, is of dimension equal to r(**A**), the rank of **A**.

Consider, for example, the set of three-dimensional vectors

$$\mathbf{a} = \begin{pmatrix} 3 \\ 0 \\ 4 \end{pmatrix}, \quad \mathbf{b} = \begin{pmatrix} 4 \\ 0 \\ 4 \end{pmatrix}, \quad \mathbf{c} = \begin{pmatrix} 4 \\ 0 \\ 2 \end{pmatrix}, \quad \text{and} \quad \mathbf{d} = \begin{pmatrix} 4 \\ 2 \\ 0 \end{pmatrix}.$$

These vectors are drawn in Figure 4.11, where the first, second, and third elements correspond to the x-axis, y-axis, and z-axis, respectively. It can be seen that the set of vectors **a**, **b**, and **c** spans only a two-dimensional space (the plane represented by the (x, z)-axes in Figure 4.11). If we form a matrix

$$\mathbf{A} = \begin{pmatrix} 3 & 4 & 4 \\ 0 & 0 & 0 \\ 4 & 4 & 2 \end{pmatrix}$$

which contains the vectors **a**, **b**, and **c** as columns, the reader can verify that $r(\mathbf{A}) = 2$. Thus, the set is linearly dependent and the reader can verify that any one vector can be expressed as a linear combination of the other two. On the other hand, the set of vectors **a**, **b**, and **d**, for example, spans a three-dimensional space (the *hyper-plane* represented by the (x, y, z)-axes in Figure 4.11). The reader can verify that for

$$\mathbf{B} = \begin{pmatrix} 3 & 4 & 4 \\ 0 & 0 & 2 \\ 4 & 4 & 0 \end{pmatrix},$$

which is the matrix containing the vectors **a**, **b**, and **d** as columns, we have $r(\mathbf{B}) = 3$, and therefore this set of vectors is linearly independent.

We end this section with the following remark. An n-dimensional space $\Re^n$ can be partitioned into subspaces each of dimension less than or equal to n. Some of the vectors in $\Re^n$ may lie entirely in a subspace of $\Re^n$. Consider, for example, the following vectors:

$$\mathbf{v}_1 = \begin{pmatrix} 2 \\ 3 \\ 0 \end{pmatrix}, \quad \mathbf{v}_2 = \begin{pmatrix} 1 \\ 0 \\ 1 \end{pmatrix}, \quad \text{and} \quad \mathbf{v}_3 = \begin{pmatrix} 0 \\ 4 \\ 0 \end{pmatrix}. \tag{4.6}$$

These vectors are linearly independent, hence they generate $\Re^3$. However, the first vector can be expressed as a linear combination of two linearly independent vectors such as the unit vectors $\mathbf{u}_1$ and $\mathbf{u}_2$ in $\Re^3$:

$$\mathbf{v}_1 = 2 \begin{pmatrix} 1 \\ 0 \\ 0 \end{pmatrix} + 3 \begin{pmatrix} 0 \\ 1 \\ 0 \end{pmatrix} = 2\,\mathbf{u}_1 + 3\,\mathbf{u}_2.$$ [3]

---

3 Recall that the ith unit vector $\mathbf{u}_i$ is a vector with 1 in the ith position and 0 elsewhere.

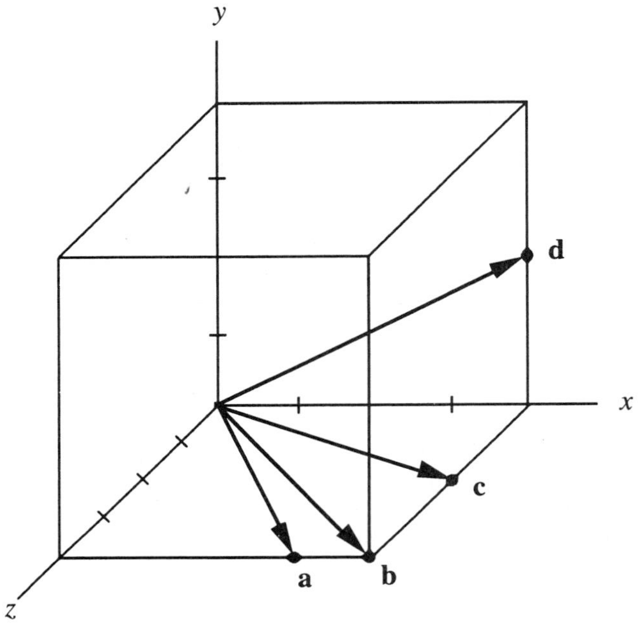

Figure 4.11. The vectors **a**, **b** and **c** are linearly dependent. They span only a two-dimensional space (the x-z plane). Any two of the vectors **a**, **b** and **c** together with the vector **d** form a set of three linearly independent vectors, hence they span a space of dimension 3.

Because it requires only two linearly independent vectors to generate $\mathbf{v}_1$, $\mathbf{v}_1$ lies in $\Re^2$. Because the third element (the z-coordinate) of $\mathbf{v}_1$ is 0, the $\Re^2$ space in which $\mathbf{v}_1$ lies is the plane formed by the x-y coordinate axes. This is the space spanned by $\mathbf{u}_1$ (which lies in the x-axis) and $\mathbf{u}_2$ (which lies in the y-axis). Similarly, the reader can verify that $\mathbf{v}_2 = \mathbf{u}_1 + \mathbf{u}_3$. Hence, $\mathbf{v}_2$ lies in a two-dimensional space, the plane formed by the x-z axes, again because the second element (the y-coordinate) of $\mathbf{v}_2$ is 0. Finally, $\mathbf{v}_3 = 4\,\mathbf{u}_2$, so the vector $\mathbf{v}_3$ lies in one-dimensional space (the y-axis in this case; why?).

As previously mentioned, a vector space $\Re^n$ can be partitioned into subspaces. This leads to the following definitions:

Definitions.

1. Subspace: A vector space $\Re^k$ is a *subspace* of $\Re^n$ (written as $\Re^k \subset \Re^n$) if every vector in $\Re^k$ is also in $\Re^n$.

2. Orthogonal subspaces: Let $\Re^k$ and $\Re^p$ be two subspaces in $\Re^n$; then $\Re^k$ and $\Re^p$ are said to be *orthogonal subspaces* (written as $\Re^k \perp \Re^p$) if (a) they

intersect only at the origin and (b) every vector in one subspace is orthogonal to every vector in the other subspace.

3. Orthogonal complement: One subspace in $\Re^n$ is said to be an *orthogonal complement* of another if (a) the two subspaces are orthogonal and (b) their union (written as $\Re^k \cup \Re^p$) is $\Re^n$.

Using a more concise notation, the last two definitions can be written as follows: Denote the intersection of two spaces $\Re^k$ and $\Re^p$ by $\Re^k \cap \Re^p$. Thus, $\Re^k \perp \Re^p$ if both of the following two conditions hold: (a) $\Re^k \cap \Re^p = \mathbf{0}$, and (b) $\mathbf{v}^T\mathbf{w} = 0$, for every vector $\mathbf{v}$ in $\Re^k$ and every vector $\mathbf{w}$ in $\Re^p$. Also, $\Re^k$ and $\Re^p$ are orthogonal complements of one another if both of the following two conditions hold: (a) $\Re^k \perp \Re^p$, and (b) $\Re^k \cup \Re^p = \Re^n$.

We conclude with a few examples using $\mathbf{v}_1$, $\mathbf{v}_2$, and $\mathbf{v}_3$ from (4.6). The following are subspaces in $\Re^3$:
(a) the $\Re^1$ space formed by the y-axis (the subspace in which $\mathbf{v}_3$ lies),
(b) the $\Re^2$ space formed by the x-z axes (the subspace in which $\mathbf{v}_2$ lies),
(c) the $\Re^2$ space formed by the x-y axes (the subspace in which $\mathbf{v}_1$ lies), and
(d) the $\Re^2$ space formed by the y-z axes.

Clearly, the first subspace is orthogonal to the second but not orthogonal to the third or to the fourth (in fact the first is a subspace of both the third and the fourth). Also, no two of the last three subspaces are pairwise orthogonal.

It should also be noted here that a space can be partitioned otherwise than in terms of the original (conventional) coordinate axes; in fact, it can be partitioned in terms of any choice of coordinate axes. Furthermore, the chosen set of coordinate axes does not have to be an orthogonal set of axes. This point will become clearer in Chapter 7, where we discuss linear transformations.

Exercises

4.1. Consider the following vectors:

$$\mathbf{a} = \begin{pmatrix} 5 \\ 1 \end{pmatrix}, \quad \mathbf{b} = \begin{pmatrix} 1 \\ -5 \end{pmatrix}, \quad \text{and} \quad \mathbf{c} = \frac{1}{2}\begin{pmatrix} 1+5\sqrt{3} \\ \sqrt{3}-5 \end{pmatrix} \cong \begin{pmatrix} 4.830 \\ -1.634 \end{pmatrix},$$

where $\mathbf{x} \cong \mathbf{y}$ means $\mathbf{x}$ is approximately equal to $\mathbf{y}$.
(a) Represent each vector graphically as a point in $\Re^2$.
(b) Represent each vector geometrically on the same graph.
(c) Compute the magnitude and direction of each vector.
(d) Determine whether each of the vectors is a normal vector.
(e) Compute the angle between each pair of vectors.
(f) Determine whether each pair of vectors is linearly independent.
(g) Determine whether each pair of vectors is orthogonal.
(h) Normalize each vector, if it is not already normal.

4.2. Consider the following vectors:

$$\mathbf{a} = \begin{pmatrix} 2 \\ 2 \\ 0 \end{pmatrix}, \quad \mathbf{b} = \begin{pmatrix} 1/\sqrt{2} \\ -1/\sqrt{2} \\ 0 \end{pmatrix}, \quad \text{and} \quad \mathbf{c} = \begin{pmatrix} 1/\sqrt{2} \\ 0 \\ 1/\sqrt{2} \end{pmatrix}.$$

(a) Compute the length of each vector.
(b) Determine whether each of the vectors is a normal vector.
(c) Compute the angle between each pair of vectors.
(d) Determine whether each pair of vectors is linearly independent.
(e) Determine whether each pair of vectors is orthogonal.
(f) Normalize each vector, if it is not already normal.

4.3. Let θ be the angle between two vectors. Determine whether the vectors are linearly independent and/or orthogonal in each of the following cases:
(a) $\theta = 0^{\circ}$. (b) $\theta = 25^{\circ}$. (c) $\theta = 90^{\circ}$.
(d) $\theta = 180^{\circ}$. (e) $\cos\theta = 0$. (f) $\cos\theta = 0.5$.
(g) $\cos\theta = 1.0$. (h) $\cos\theta = -0.5$. (i) $\cos\theta = -1.0$.

4.4. Let $\mathbf{a}^{\mathrm{T}} = (4 \quad 1)$. Find the normal vector $\mathbf{z}$ that makes an angle of α with $\mathbf{a}$, where
(a) $\alpha = 30^{\circ}$. (b) $\alpha = 60^{\circ}$. (c) $\alpha = 90^{\circ}$.
[Hint: Use the matrix **A** in Exercise 2.7 and normalize.]

4.5. The length of a vector $\mathbf{v}$ is 4. What are the algebraic and geometric effects of multiplying $\mathbf{v}$ by s, where:
(a) $s = 0$. (b) $s = 1$. (c) $s = 2$. (d) $s = -2$. (e) $s = \|\mathbf{v}\|$.

4.6. The norms of two vectors $\mathbf{u}$ and $\mathbf{v}$ of the same order are 2 and $\sqrt{2}$, respectively. Their inner product is 2. The first is multiplied by a and the second by b. The resultant vectors are $\mathbf{x}$ and $\mathbf{y}$, respectively.
(a) Compute the lengths of $\mathbf{x}$ and $\mathbf{y}$.
(b) Compute the inner product of $\mathbf{x}$ and $\mathbf{y}$.
(c) Compute the angle between $\mathbf{u}$ and $\mathbf{v}$.
(d) Compute the angle between $\mathbf{x}$ and $\mathbf{y}$.
(e) Compute the angle between $\mathbf{u}$ and $\mathbf{x}$.
(f) Compute the angle between $\mathbf{u}$ and $\mathbf{y}$.
(g) Are $\mathbf{u}$ and $\mathbf{v}$ orthogonal? Are they linearly independent?
(h) Are $\mathbf{u}$ and $\mathbf{x}$ orthogonal? Are they linearly independent?

4.7. Consider the vectors $\mathbf{a}^{\mathrm{T}} = (4 \quad 1)$ and $\mathbf{b}^{\mathrm{T}} = (0 \quad 3)$.
(a) Compute $\mathbf{c} = \mathbf{a} + \mathbf{b}$ and $\mathbf{d} = \mathbf{a} - \mathbf{b}$.
(b) Draw the vectors **a**, **b**, and **c** in a graph that illustrates the *parallelogram rule* for vector addition.

(c) Draw the vectors **a**, **b**, and **d** in a graph that illustrates the *parallelogram rule* for vector subtraction.

4.8. If **x**, **y**, and **z** are linearly independent, show that the vectors $(\mathbf{x}+\mathbf{y})$, $(\mathbf{x}+\mathbf{z})$, and $(\mathbf{y}+\mathbf{z})$ are also linearly independent.

4.9. Express $(3 \quad -6)^T$ as a linear combination of each of the following sets:

(a) $\mathbf{v}_1 = \begin{pmatrix} 0 \\ 1 \end{pmatrix}, \mathbf{v}_2 = \begin{pmatrix} 1 \\ 0 \end{pmatrix}.$ (b) $\mathbf{v}_1 = \begin{pmatrix} 1 \\ 0 \end{pmatrix}, \mathbf{v}_2 = \begin{pmatrix} 1 \\ 2 \end{pmatrix}.$

(c) $\mathbf{v}_1 = \begin{pmatrix} -1 \\ 2 \end{pmatrix}, \mathbf{v}_2 = \begin{pmatrix} 2 \\ 1 \end{pmatrix}.$ (d) $\mathbf{v}_1 = \begin{pmatrix} 3 \\ 3 \end{pmatrix}, \mathbf{v}_2 = \begin{pmatrix} 2 \\ 1 \end{pmatrix}, \mathbf{v}_3 = \begin{pmatrix} 1 \\ 2 \end{pmatrix}.$

4.10. Let $\mathbf{y} = \mathbf{A}\mathbf{x}$, where

$$\mathbf{A} = \begin{pmatrix} 1 & 1 \\ 1 & -1 \end{pmatrix} \quad \text{and} \quad \mathbf{x} = \begin{pmatrix} 3 \\ 2 \end{pmatrix}.$$

(a) Graph the vectors **x** and **y**.
(b) Compute the magnitudes of **x** and **y**.
(c) Do **x** and **y** have the same direction?

4.11. If possible, express $(5 \quad 2)^T$ as a linear combination of each of the following sets of vectors; if not, indicate the reason.

(a) $\mathbf{v}_1 = \begin{pmatrix} 0 \\ 1 \end{pmatrix}, \mathbf{v}_2 = \begin{pmatrix} 1 \\ 0 \end{pmatrix}.$ (b) $\mathbf{v}_1 = \begin{pmatrix} 1 \\ 0 \end{pmatrix}, \mathbf{v}_2 = \begin{pmatrix} 1 \\ 2 \end{pmatrix}.$

(c) $\mathbf{v}_1 = \begin{pmatrix} 1 \\ 0 \end{pmatrix}, \mathbf{v}_2 = \begin{pmatrix} 2 \\ 0 \end{pmatrix}.$ (d) $\mathbf{v}_1 = \begin{pmatrix} 1 \\ 0 \end{pmatrix}, \mathbf{v}_2 = \begin{pmatrix} 0 \\ 0 \end{pmatrix}.$

(e) $\mathbf{v}_1 = \begin{pmatrix} 1 \\ 2 \end{pmatrix}, \mathbf{v}_2 = \begin{pmatrix} 2 \\ 1 \end{pmatrix}, \mathbf{v}_3 = \begin{pmatrix} 2 \\ 4 \end{pmatrix}.$ (f) $\mathbf{v}_1 = \begin{pmatrix} 2 \\ 1 \end{pmatrix}, \mathbf{v}_2 = \begin{pmatrix} 4 \\ 2 \end{pmatrix}, \mathbf{v}_3 = \begin{pmatrix} 6 \\ 3 \end{pmatrix}.$

4.12. State whether each of the sets of vectors in Exercise 4.11 is a spanning set for $\Re^2$.

4.13. State whether each of the sets in Exercise 4.11 forms a basis for $\Re^2$.

4.14. What is the dimension of the space spanned by each set in Exercise 4.11?

4.15. Graph the vectors $\mathbf{v}_1$, $\mathbf{v}_2$, and $\mathbf{v}_3$ in (4.6) using a three-dimensional space as in Figure 4.11 and check that:
(a) $\mathbf{v}_1$ lies in the plane represented by the x-y axes.
(b) $\mathbf{v}_2$ also lies in a plane (which one?).
(c) $\mathbf{v}_3$ also lies in a one-dimensional subspace (which one?).

4.16. In each of the following cases, consider the space spanned by the columns of $\mathbf{V}$ and that spanned by the columns of $\mathbf{W}$. What is the dimension of each of these spaces? Also, state whether one space is a subspace of the other and whether it is orthogonal to the other.

(a) $\mathbf{V} = \begin{pmatrix} 1 & 0 \\ 0 & 0 \\ 0 & 1 \end{pmatrix}, \mathbf{W} = \begin{pmatrix} 2 & 0 \\ 0 & 0 \\ 0 & 3 \end{pmatrix}.$ (b) $\mathbf{V} = \mathbf{I}_3, \mathbf{W} = \begin{pmatrix} 2 & 0 & 1 \\ 0 & 0 & 1 \\ 0 & 1 & 0 \end{pmatrix}.$

(c) $\mathbf{V} = \mathbf{I}_3, \mathbf{W} = \begin{pmatrix} 0 & 0 \\ 1 & 0 \\ 0 & 1 \end{pmatrix}.$ (d) $\mathbf{V} = \mathbf{I}_3, \mathbf{W} = \begin{pmatrix} 2 & 0 & 1 \\ 0 & 0 & 0 \\ 0 & 1 & 0 \end{pmatrix}.$

5 Three Matrix Reductions

In many applications it is useful to summarize or characterize a given matrix by a single number. We refer to this as matrix reduction. In this chapter, we define and discuss three such reductions: trace, determinant, and norms.

5.1. Trace

Definition. The *trace* of an $n \times n$ matrix $\mathbf{A} = (a_{ij})$, denoted by $\text{tr}(\mathbf{A})$, is the sum of the diagonal elements of $\mathbf{A}$, namely,

$$\text{tr}(\mathbf{A}) = \sum_{i=1}^{n} a_{ii}\,. \tag{5.1}$$

Note that the trace is defined only for square matrices. For example, the matrices

$$\mathbf{A} = \begin{pmatrix} 2 & 1 & 4 \\ 0 & 1 & 2 \\ 0 & 0 & 0 \end{pmatrix}, \quad \mathbf{B} = \begin{pmatrix} 0 & 2 & 3 & 0 \\ 0 & 0 & 1 & 5 \\ 0 & 0 & 0 & 0 \\ 0 & 0 & 0 & 0 \end{pmatrix}, \quad \mathbf{C} = \begin{pmatrix} 0 & 2 & -4 & 6 \\ 1 & 2 & 3 & 0 \\ 0 & -1 & 2 & 4 \\ -2 & -4 & -9 & 1 \end{pmatrix}$$

have $\text{tr}(\mathbf{A}) = 3$, $\text{tr}(\mathbf{B}) = 0$, and $\text{tr}(\mathbf{C}) = 5$. It also follows from the definition of the trace that for any scalar (1×1 matrix) s, $\text{tr}(s) = s$.

The trace of a matrix has an interesting interpretation. From (5.1) we can write

$$\text{tr}(\mathbf{A}) = n\left(\frac{1}{n}\sum_{i=1}^{n} a_{ii}\right) = n\bar{a}_{ii},$$

where $\bar{a}_{ii}$ is the *arithmetic mean* (average) of the diagonal elements of $\mathbf{A}$. Thus, $\text{tr}(\mathbf{A})$ is equal to n times the average diagonal element of $\mathbf{A}$. Other properties of the trace are given below.

Result 5.1. The trace of a matrix has the following properties:
(a) $\text{tr}(\mathbf{A}^{\text{T}}) = \text{tr}(\mathbf{A})$.
(b) $\text{tr}(s\mathbf{A}) = s\,\text{tr}(\mathbf{A})$, where s is a scalar.
(c) $\text{tr}(\mathbf{A} \pm \mathbf{B}) = \text{tr}(\mathbf{A}) \pm \text{tr}(\mathbf{B})$, provided that $\mathbf{A}$ and $\mathbf{B}$ are square matrices of the same order.

(d) $\text{tr}(\mathbf{ABC}) = \text{tr}(\mathbf{BCA}) = \text{tr}(\mathbf{CAB})$, provided the products exist.
(e) If $\mathbf{A}$ is an idempotent matrix, then $\text{r}(\mathbf{A}) = \text{tr}(\mathbf{A})$.[1]
(f) $\text{tr}(\mathbf{A}\otimes\mathbf{B}) = \text{tr}(\mathbf{A})\,\text{tr}(\mathbf{B})$.

Result 5.1(d) is known as the *cyclical* property of the trace. This property is useful, for example, in simplifying several formulas in regression analysis, so we illustrate it by an example. Consider the vectors

$$\mathbf{a} = \begin{pmatrix} 2 \\ 4 \\ 1 \end{pmatrix} \quad \text{and} \quad \mathbf{b} = \begin{pmatrix} 3 \\ 6 \\ 2 \end{pmatrix}.$$

Using Result 5.1(d), we have $\text{tr}(\mathbf{a}^{\mathrm{T}}\mathbf{b}) = \text{tr}(\mathbf{b}\mathbf{a}^{\mathrm{T}})$. This can be verified as follows. Because $\mathbf{a}^{\mathrm{T}}\mathbf{b}$ is a scalar, $\text{tr}(\mathbf{a}^{\mathrm{T}}\mathbf{b}) = \mathbf{a}^{\mathrm{T}}\mathbf{b} = 32$. On the other hand,

$$\mathbf{b}\mathbf{a}^{\mathrm{T}} = \begin{pmatrix} 3 \\ 6 \\ 2 \end{pmatrix} \begin{pmatrix} 2 & 4 & 1 \end{pmatrix} = \begin{pmatrix} 6 & 12 & 3 \\ 12 & 24 & 6 \\ 4 & 8 & 2 \end{pmatrix}$$

is a matrix and $\text{tr}(\mathbf{b}\mathbf{a}^{\mathrm{T}}) = 32$.

Result 5.1(e) gives an easy way to compute the rank of idempotent matrices. For example, let $\mathbf{1}$ be the vector of n ones; then $\mathbf{A} = n^{-1}\mathbf{1}\mathbf{1}^{\mathrm{T}}$ is an idempotent matrix. Each of its elements is $1/n$. Thus, by Result 5.1(e), $\text{r}(\mathbf{A}) = \text{tr}(\mathbf{A}) = 1$. For $n = 3$, for example, the reader can verify that the matrix

$$\mathbf{A} = \frac{1}{3}\begin{pmatrix} 1 & 1 & 1 \\ 1 & 1 & 1 \\ 1 & 1 & 1 \end{pmatrix}$$

is an idempotent matrix. (Compute a row-echelon form of $\mathbf{A}$ and check that it contains one nonzero row and that therefore $\text{r}(\mathbf{A}) = 1$.)

Although the only idempotent scalars are 0 and 1, idempotent matrices form an uncountable set. Idempotent matrices play an important role in statistics and we shall encounter them in several places in this book (e.g., Sections 6.3, 7.5, and 8.3).

Finally, Result 5.1(f) asserts that the trace of a Kronecker product is the product of the traces.

5.2. Determinant

Like the trace, the determinant is defined only for square matrices. But the definition of the determinant for a square matrix of any order is not as easy as that of the trace. We postpone the definition of the determinant of a general matrix until Chapter 9 because the topics introduced there give rise to an easy definition of the determinant. In this section, we define the determinants of 1×1 and 2×2 matrices. We then give

[1] Recall that $\mathbf{A}$ is called an idempotent matrix if $\mathbf{A}^2 = \mathbf{A}$.

some examples and define the determinant of some special matrices. We also state some properties of the determinant. These properties are the same regardless of the order of the matrix, and they give rise to a method for computing the determinant of a square matrix of any order. This method is also presented in this section.

Definitions.
(a) If $\mathbf{A} = (a_{11})$ (i.e., a scalar), then the determinant of $\mathbf{A}$, denoted by $\det(\mathbf{A})$, is simply a_{11}, the sole element of $\mathbf{A}$.

(b) For any 2×2 matrix $\mathbf{A} = \begin{pmatrix} a_{11} & a_{12} \\ a_{21} & a_{22} \end{pmatrix}$, the determinant of $\mathbf{A}$ is defined by

$$\det(\mathbf{A}) = a_{11}\, a_{22} - a_{21}\, a_{12}. \tag{5.2}$$

In words, the determinant of a 2×2 matrix is equal to the difference between the product of the two major diagonal elements and the product of the two minor diagonal elements. To illustrate, let

$$\mathbf{A} = \begin{pmatrix} 5 & -2 \\ 3 & 4 \end{pmatrix} \quad \text{and} \quad \mathbf{B} = \begin{pmatrix} 3 & 1 \\ 6 & 2 \end{pmatrix}; \tag{5.3}$$

then $\det(\mathbf{A}) = 5 \times 4 - 3 \times (-2) = 26$ and $\det(\mathbf{B}) = 3 \times 2 - 6 \times 1 = 0$.

The determinants of some special matrices are given in the following result.

Result 5.2. Let $\mathbf{A}$ be a square matrix of any order.
(a) If $\mathbf{A}$ is a diagonal or a triangular matrix, then $\det(\mathbf{A})$ is the product of the diagonal elements.
(b) If $\mathbf{A}$ is an orthogonal matrix, then $\det(\mathbf{A})$ is either 1 or -1.

As can be seen from Result 5.2(a), the determinants of diagonal and triangular matrices are easy to compute. Furthermore, the determinant of a diagonal matrix has an interesting interpretation. First note that the nth root of the product of n numbers is called the *geometric mean* of the n numbers. Now, from Result 5.2(a), if $\mathbf{A}$ is diagonal or triangular, then $\det(\mathbf{A})$ can be written as

$$\det(\mathbf{A}) = a_{11} \times a_{22} \times \cdots \times a_{nn} = \left(\sqrt[n]{a_{11} \times a_{22} \times \cdots \times a_{nn}}\right)^n = \tilde{a}_{ii}^n,$$

where $\tilde{a}_{ii}$ denotes the geometric mean of the diagonal elements of $\mathbf{A}$. Thus, the determinant of a diagonal or a triangular matrix is simply the geometric mean of its diagonal elements raised to the nth power.

Result 5.2(a) also suggests a strategy for computing the determinant. We know from Chapter 3 that we can always transform any matrix to an upper-triangular matrix (row-echelon form). If we can do so without changing the value of the determinant, we can then compute the determinant easily. The question is then, What are the effects of the various types of row operations on the determinant of a matrix? The answer is given in the following result.

Result 5.3. The effects of the elementary row operations on the determinant of a matrix are:

(a) Operations of Type 1: Interchanging two rows (or columns) of a matrix changes the algebraic sign of its determinant.
(b) Operations of Type 2: When a row (or column) of a matrix is multiplied by a scalar s, the determinant of the matrix is multiplied by s.
(c) Operations of Type 3: Adding to a row a nonzero multiple of another row of a matrix does not change the determinant.

To illustrate, let

$$\mathbf{A} = \begin{pmatrix} 4 & 1 \\ 2 & 2 \end{pmatrix};$$

then $\det(\mathbf{A}) = (4 \times 2) - (2 \times 1) = 6$. Interchanging the rows of $\mathbf{A}$ gives

$$\mathbf{A}_1 = \begin{pmatrix} 2 & 2 \\ 4 & 1 \end{pmatrix},$$

with $\det(\mathbf{A}_1) = -6 = -\det(\mathbf{A})$. When we interchange the columns of $\mathbf{A}_1$, we obtain

$$\mathbf{A}_2 = \begin{pmatrix} 2 & 2 \\ 1 & 4 \end{pmatrix},$$

with $\det(\mathbf{A}_2) = 6 = \det(\mathbf{A})$. Thus, after any *even* number of interchanges of two rows or two columns, the determinant remains unchanged. Let us now multiply one row of $\mathbf{A}$, say the first, by 2. We obtain

$$\mathbf{A}_3 = \begin{pmatrix} 8 & 2 \\ 2 & 2 \end{pmatrix},$$

with $\det(\mathbf{A}_3) = 12$, which is twice the determinant of $\mathbf{A}$. Finally, if we perform a row operation of Type 3, such as multiplying row 1 of $\mathbf{A}$ by -0.5 and adding the result to row 2, we obtain

$$\mathbf{A}_4 = \begin{pmatrix} 4 & 1 \\ 0 & 1.5 \end{pmatrix},$$

with $\det(\mathbf{A}_4) = 6 = \det(\mathbf{A})$. Thus, row operations of Type 3 do not alter the determinant. Note that $\mathbf{A}_4$ is a triangular matrix and its determinant can also be obtained by multiplying the diagonal elements as Result 5.2(a) indicates.

Results 5.2(a) and 5.3 set the stage for a method of computing the determinant of any square matrix. This method can be summarized as follows:

1. If necessary, reduce the matrix to an upper-triangular matrix using only row operations of Types 1 and 3.

2. If the number of row operations of Type 1 is even, then the determinant is simply the product of the diagonal elements of the upper-triangular matrix; otherwise, the determinant is the negative of the product of the diagonal elements of the upper-triangular matrix.

As an example, to compute the determinant of

$$\mathbf{A} = \begin{pmatrix} 2 & 6 & 4 \\ 5 & 30 & 15 \\ 6 & 3 & 9 \end{pmatrix}, \tag{5.4}$$

we first need to reduce $\mathbf{A}$ to an upper-triangular form using only row operations of Types 1 and 3. The reader can verify that, using only operations of Type 3, $\mathbf{A}$ can be reduced to

$$\mathbf{A}_1 = \begin{pmatrix} 2 & 6 & 4 \\ 0 & 15 & 5 \\ 0 & 0 & 2 \end{pmatrix}. \tag{5.5}$$

Since row operations of Type 3 have no effect on the determinant, then $\det(\mathbf{A}) = \det(\mathbf{A}_1) = 2 \times 15 \times 2 = 60$.

The determinant has several other useful and interesting properties, some of which are stated in the following result.

Result 5.4. The determinant has the following properties:

(a) $\det(\mathbf{A}^{\mathrm{T}}) = \det(\mathbf{A})$.

(b) $\det(\mathbf{A}) = 0$ if and only if $\mathrm{r}(\mathbf{A}) < n$, where $\mathbf{A}$ is $n \times n$.

(c) $\det(s\mathbf{A}) = s^n \det(\mathbf{A})$, where s is a scalar and $\mathbf{A}$ is $n \times n$.

(d) $\det(\mathbf{A}^{\mathrm{T}}\mathbf{A}) \neq 0$ if and only if $\mathrm{r}(\mathbf{A}) = n$, where $\mathbf{A}$ is $m \times n$.

(e) $\det(\mathbf{AB}) = \det(\mathbf{A}) \det(\mathbf{B})$, provided that $\mathbf{A}$ and $\mathbf{B}$ are square and of the same order.

(f) If $\det(\mathbf{AB}) = 0$, then either $\det(\mathbf{A}) = 0$ or $\det(\mathbf{B}) = 0$ or both, provided that $\mathbf{A}$ and $\mathbf{B}$ are square and of the same order.

(g) If $m > n$, then $\det(\mathbf{A}_{m \times n} \mathbf{B}_{n \times m}) = 0$.

(h) $\det(\mathbf{A} \otimes \mathbf{B}) = (\det(\mathbf{A}))^m (\det(\mathbf{B}))^n$, where $\mathbf{A}$ is $n \times n$ and $\mathbf{B}$ is $m \times m$.

Part (a) states that transposing a matrix leaves its determinant unchanged. Part (b) is of particular interest because it shows the relationships among the determinant, the rank, and hence the linear dependency of vectors in square matrices. It states that if the determinant of $\mathbf{A}$ is 0, then $\mathbf{A}$ must be rank-deficient (i.e., the rows (and columns) of $\mathbf{A}$ must be linearly dependent). The converse is also true: If $\mathbf{A}$ is rank-deficient, the determinant of $\mathbf{A}$ must be 0. Part (c) follows from Result 5.3(b) because multiplying a matrix by a scalar s is equivalent to multiplying each row of the matrix by s. Part (d) asserts that $\mathbf{A}^{\mathrm{T}}\mathbf{A}$ has a nonzero determinant if and only if $\mathbf{A}$ is of full column rank. It follows from Part (d) that $\mathbf{A}\mathbf{A}^{\mathrm{T}}$ has a nonzero determinant if and only if $\mathbf{A}$ is of full row rank. Part (e) states that the determinant of a product of square matrices is the product of the determinants of the individual matrices. The remaining parts are self-explanatory. Result 5.4(b) leads to the following definition.

Definition. A square matrix with a zero determinant is called a *singular* matrix, and a square matrix with a nonzero determinant is called a *nonsingular* matrix.

Thus, linear dependency, rank-deficiency, and singularity amount to the same thing. For example, consider the matrices

$$\mathbf{A} = \begin{pmatrix} 5 & -2 \\ 3 & 4 \end{pmatrix} \quad \text{and} \quad \mathbf{B} = \begin{pmatrix} 3 & 1 \\ 6 & 2 \end{pmatrix}.$$

Since $\det(\mathbf{A}) \neq 0$, $\mathbf{A}$ must be of full rank, that is, the two rows (columns) of $\mathbf{A}$ are linearly independent. On the other hand, since $\det(\mathbf{B}) = 0$, $\mathbf{B}$ must be rank-deficient, that is, the two vectors (rows or columns) in $\mathbf{B}$ are linearly dependent (the first column is three times the second and the first row is half the second).

Another useful but not obvious property of the determinant is that it is related to the volume of certain geometric shapes in multidimensional spaces. For example, it is related to the volume of an ellipsoid (see Result 10.2(e)), the volume of a parallelepiped (see, e.g., Strang, 1988), and the area of a triangle (see Exercise 5.18). Other properties of the determinant are given in Chapter 6, after matrix inversion is defined (see, for example, Results 6.2, 6.6, and 6.7).

5.3. Vector Norms

The magnitude of a scalar a is measured by quantities such as $|a|$, the absolute value of a, and a^2. In Section 4.3 we identified the magnitude of a vector with the length of its geometric representation. In this section we present other ways of measuring the magnitude of a vector. Measuring the magnitude of a matrix is postponed until Section 11.1 because it requires concepts to be introduced in Chapter 9.

The magnitude of a vector can be measured in terms of the magnitudes of its elements. One of the most common measures of the magnitude of a vector

$$\mathbf{v} = \begin{pmatrix} v_1 \\ v_2 \\ \vdots \\ v_n \end{pmatrix}$$

is given by

$$\|\mathbf{v}\|_p = \{|v_1|^p + |v_2|^p + \cdots + |v_n|^p\}^{1/p} \tag{5.6}$$

where $p \geq 1$. This is known as the *p-norm* or the L_p*-norm* and is denoted by $\|\mathbf{v}\|_p$. To compute the L_p-norm we need to raise the absolute value of each element to the power p, sum the results, and then take the (positive) pth root of the sum. The most commonly used values of p are 1, 2, and infinity. In these cases, (5.6) becomes

$$\|\mathbf{v}\|_1 = |v_1| + |v_2| + \cdots + |v_n|, \tag{5.7}$$

$$\|\mathbf{v}\|_2 = \{|v_1|^2 + |v_2|^2 + \cdots + |v_n|^2\}^{1/2}, \tag{5.8}$$

$$\|\mathbf{v}\|_\infty = \max\{|v_1|, |v_2|, \ldots, |v_n|\}, \tag{5.9}$$

respectively. In words, $\|\mathbf{v}\|_1$ is simply the sum of the absolute values of the elements, $\|\mathbf{v}\|_2$ is the square root of the sum of the squared elements, and $\|\mathbf{v}\|_\infty$ is the maximum absolute value of the elements. As an example, for the vectors

$$\mathbf{a} = \begin{pmatrix} 2 \\ 1 \\ -2 \end{pmatrix} \quad \text{and} \quad \mathbf{b} = \begin{pmatrix} 0 \\ 0 \\ 4 \end{pmatrix}, \tag{5.10}$$

we have

$$\|\mathbf{a}\|_1 = |2| + |1| + |-2| = 5,$$
$$\|\mathbf{b}\|_1 = |0| + |0| + |4| = 4,$$

$$\|\mathbf{a}\|_2 = \{|2|^2 + |1|^2 + |-2|^2\}^{1/2} = 3,$$
$$\|\mathbf{b}\|_2 = \{|0|^2 + |0|^2 + |4|^2\}^{1/2} = 4,$$

$$\|\mathbf{a}\|_\infty = \max\{|2|, |1|, |-2|\} = 2,$$
$$\|\mathbf{b}\|_\infty = \max\{|0|, |0|, |4|\} = 4.$$

If we use the L_1-norm to measure the magnitude of vectors, the vector **a** is larger than the vector **b**. However, if we use either the L_2-norm or the L_∞-norm, **b** is larger than **a**. Thus, different norms can produce different orderings of vectors.

Of the three norms above, the L_2-norm is the most common. For this reason and for notational simplicity, we use $\|\mathbf{v}\|$ instead of $\|\mathbf{v}\|_2$ to denote the L_2-norm of **v**. Also, when we say simply "the norm of **v**," we mean the L_2-norm.

From the definition of the norm in (5.8), we see that the norm of **v** is nothing but the square root of the inner product of the vector and itself; that is,

$$\|\mathbf{v}\| = \sqrt{v_1^2 + v_2^2 + \cdots + v_n^2} = \sqrt{\mathbf{v}^{\mathrm{T}}\mathbf{v}}\,. \tag{5.11}$$

This is the same as formula (4.2), which was used to define the length of the line segment representing the vector in an n-dimensional space. Vector norms have several interesting properties, some of which are stated in the next result.

Result 5.5. The L_p-norms have the following properties:
(a) $\|\mathbf{0}\|_p = 0$, where $\mathbf{0}$ is the null vector.
(b) $\|\mathbf{v}\|_p > 0$ for $\mathbf{v} \neq \mathbf{0}$.
(c) $\|s\mathbf{v}\|_p = |s| \times \|\mathbf{v}\|_p$ for any scalar s.
(d) $\|\mathbf{u} + \mathbf{v}\|_p \leq \|\mathbf{u}\|_p + \|\mathbf{v}\|_p$.

(e) $|\mathbf{u}^{\mathrm{T}}\mathbf{v}| \leq \|\mathbf{u}\|_p \times \|\mathbf{v}\|_p$.
(f) $\|\mathbf{v}\|_\infty \leq \|\mathbf{v}\|_1 \leq n\,\|\mathbf{v}\|_\infty$, where n is the number of elements in $\mathbf{v}$.
(g) $\|\mathbf{v}\|_\infty \leq \|\mathbf{v}\|_2 \leq \sqrt{n}\,\|\mathbf{v}\|_\infty$, where n is the number of elements in $\mathbf{v}$.

Parts (a) and (b) indicate that the L_p-norm is a nonnegative number. Part (c) states that the L_p-norm of a vector multiplied by a scalar is equal to the L_p-norm of the vector multiplied by the absolute value of the scalar. Part (d) is called the *triangle inequality* because it is a generalized version of the fact that the length of any side of a triangle is less than or equal to the sum of the lengths of the other two sides. To illustrate this property, we use the vectors in (5.10). We compute

$$\mathbf{a}+\mathbf{b} = \begin{pmatrix} 2 \\ 1 \\ 2 \end{pmatrix},$$

hence $\|\mathbf{a}+\mathbf{b}\| = \sqrt{4+1+4} = 3$, which is less than $\|\mathbf{a}\| + \|\mathbf{b}\| = 7$.

Part (e) states that the absolute value of the inner product of two vectors is less than or equal to the product of the L_p-norms of the individual vectors. When $p = 2$, this result is known as the *Cauchy-Schwarz inequality*. Thus, for example, for the vectors **a** and **b** in (5.10), we have

$$\mathbf{a}^{\mathrm{T}}\mathbf{b} = \begin{pmatrix} 2 & 1 & -2 \end{pmatrix} \begin{pmatrix} 0 \\ 0 \\ 4 \end{pmatrix} = -8,$$

and

$$\|\mathbf{a}\| \times \|\mathbf{b}\| = 3 \times 4 = 12,$$

from which we see that $|\mathbf{a}^{\mathrm{T}}\mathbf{b}| < \|\mathbf{a}\| \times \|\mathbf{b}\|$.

Note that Part (e) follows directly from Result 4.1, which states that the angle θ between two vectors **u** and **v** satisfies

$$\cos\theta = \frac{\mathbf{u}^{\mathrm{T}}\mathbf{v}}{\|\mathbf{u}\| \times \|\mathbf{v}\|},$$

which implies that $\mathbf{u}^{\mathrm{T}}\mathbf{v} = \cos\theta \times \|\mathbf{u}\| \times \|\mathbf{v}\|$. Taking the absolute value of both sides, we get $|\mathbf{u}^{\mathrm{T}}\mathbf{v}| = |\cos\theta| \times \|\mathbf{u}\| \times \|\mathbf{v}\|$. But since $-1 \leq \cos\theta \leq 1$, $|\cos\theta| \leq 1$ and Result 5.5(e) immediately follows.

Finally, Parts (f) and (g) show some of the many relationships that exist among the various vector norms.

We have seen in this section that the norm measures the magnitude of vectors, which allows us, using a given norm, to order vectors according to their magnitudes. In addition, the concept of vector norms has many other practical applications. We have already seen in Chapter 4, for example, that the norm has some geometrical interpretations and is useful in measuring distances between points in a multidimensional space.

Exercises

5.1. For each of the following matrices:

$$\mathbf{A}=\begin{pmatrix}1 & 2\\ 2 & 6\end{pmatrix},\ \mathbf{B}=\begin{pmatrix}1 & 3\\ 2 & -6\end{pmatrix},\ \mathbf{C}=\begin{pmatrix}2 & 1 & 2\\ 1 & 0 & 1\end{pmatrix},\ \mathbf{D}=\begin{pmatrix}0 & 0\\ 0 & 0\\ 0 & 0\end{pmatrix},\ \mathbf{E}=\begin{pmatrix}0 & 0 & 0\\ 1 & 0 & 1\end{pmatrix}$$

(a) Compute the rank of the transpose. (b) Compute the trace, where possible.
(c) Determine whether the set of column vectors is linearly dependent.
(d) Determine whether the set of row vectors is linearly dependent.
(e) Determine whether the matrix is full-column rank, full-row rank, and/or rank-deficient.
(f) What is the dimension of the space spanned by the column vectors?

5.2. Repeat Exercise 5.1 for each of the matrices in Exercise 3.5.

5.3. Repeat Exercise 5.1 for each of the matrices in Exercise 3.8.

5.4. Given the matrices in Exercise 5.1, find the trace of the following matrices without actually computing the indicated expressions:

(a) $3\mathbf{A}$ (b) $\mathbf{A}+\mathbf{B}$ (c) $\mathbf{A}-\mathbf{B}$ (d) $\mathbf{ABCDE}$ (e) $\mathbf{A}\otimes\mathbf{C}$

5.5. Given the matrices in Exercise 5.1, find the rank of the following matrices without actually computing the indicated products:

(a) $3\mathbf{A}$ (b) $\mathbf{C}^{\mathrm{T}}\mathbf{C}$ (c) $\mathbf{CC}^{\mathrm{T}}$ (d) $\mathbf{A}\otimes\mathbf{B}$ (e) $\mathbf{A}\otimes\mathbf{C}$

5.6. Let **a**, **b**, **c**, and **d** be vectors of the same order such that $\mathbf{a}^{\mathrm{T}}\mathbf{d}=2$ and $\mathbf{b}^{\mathrm{T}}\mathbf{c}=4$. Compute the trace of $\mathbf{ab}^{\mathrm{T}}\mathbf{cd}^{\mathrm{T}}$.

5.7. Let **x** and **y** be orthonormal vectors. Compute $\mathrm{tr}(\mathbf{xx}^{\mathrm{T}})$ and $\mathrm{tr}(\mathbf{xx}^{\mathrm{T}}\mathbf{yy}^{\mathrm{T}})$.

5.8. Let **1** be a vector of n ones; then
(a) Show that $\mathbf{A}=n^{-1}\mathbf{11}^{\mathrm{T}}$ is an idempotent matrix.
(b) Using $n=3$, verify that $\mathrm{tr}(\mathbf{A})=\mathrm{r}(\mathbf{A})=3$.

5.9. Let $\mathbf{A}_{3\times 5}$, $\mathbf{B}_{3\times 5}$, and $\mathbf{C}_{5\times 5}$ be matrices of the indicated orders. Suppose that $\mathrm{r}(\mathbf{A})=2$, $\mathrm{r}(\mathbf{B})=3$, and $\mathrm{r}(\mathbf{C})=4$. State whether each of the following statements is true or false (explain):
(a) $\mathrm{r}(\mathbf{A}+\mathbf{B})\le 4$. (b) $\mathrm{r}(\mathbf{B})\le 3$. (c) $\mathrm{r}(\mathbf{BC})\le 3$.
(d) $\det(\mathbf{A})=\det(\mathbf{C})=0$. (f) $\det(\mathbf{A}^{\mathrm{T}}\mathbf{A})=0$. (g) $\det(\mathbf{C}^2)=(\det(\mathbf{C}))^2$.
(h) $\mathrm{r}(-\mathbf{A})=-\mathrm{r}(\mathbf{A})$. (i) $\mathrm{r}(\mathbf{C}+\mathbf{C}^{\mathrm{T}})=\mathrm{r}(\mathbf{C})$. (j) $\mathrm{r}(\mathbf{C}^2)=(\mathrm{r}(\mathbf{C}))^2$.
(k) There are three linearly independent rows in **A**.
(l) There are five linearly independent columns in **A**.
(m) **A** is rank-deficient, therefore it is singular.
(n) $\mathrm{r}(\mathbf{AE})=\mathrm{r}(\mathbf{A})$, where **E** is an elementary matrix.

5.10. For each of the matrices in Exercise 5.1:
(a) Compute the determinant, where possible.
(b) Classify the matrix as either singular or nonsingular, where possible.

5.11. Compute the determinant of each of the following matrices:

$$\mathbf{A} = \begin{pmatrix} 2 & 3 & 0 \\ 0 & 0 & 5 \\ 0 & 0 & 1 \end{pmatrix}, \quad \mathbf{B} = \begin{pmatrix} 4 & 0 & 0 \\ 0 & 1 & 0 \\ 0 & 0 & -1 \end{pmatrix}, \quad \mathbf{C} = \begin{pmatrix} 2 & 0 & 0 \\ 6 & 2 & 0 \\ 0 & 1 & 1 \end{pmatrix}.$$

5.12. Let $\mathbf{I}$ be the identity matrix of order 2 and $\mathbf{A}$ be a square matrix of order 2 with $\det(\mathbf{A}) = d$. Suppose that $\mathbf{E}_1$ is obtained by multiplying the first row of $\mathbf{I}$ by 2, $\mathbf{E}_2$ is obtained by interchanging the two rows of $\mathbf{I}$, $\mathbf{E}_3$ is obtained by replacing the first row of $\mathbf{I}$ by the sum of its rows, and $\mathbf{C}_1$ is obtained by interchanging the columns of $\mathbf{I}$. Compute the determinant of:

(a) $\mathbf{E}_1\mathbf{A}$ (b) $\mathbf{A}\mathbf{C}_1$ (c) $\mathbf{E}_1\mathbf{A}\mathbf{C}_1$
(d) $\mathbf{E}_2\mathbf{E}_1\mathbf{A}$ (e) $\mathbf{E}_3\mathbf{E}_2\mathbf{E}_1\mathbf{A}$ (f) $\mathbf{E}_3\mathbf{E}_2\mathbf{E}_1\mathbf{A}\mathbf{C}_1$

5.13. Use the matrix $\mathbf{A}$ in Exercise 5.1 to verify your computations in Exercise 5.12.

5.14. Let $\mathbf{A}$ be a square matrix. Show that $\det(\mathbf{A}) = 0$ if one or more of the following conditions hold:

(a) Two or more rows of $\mathbf{A}$ are equal.
(b) Two or more columns of $\mathbf{A}$ are equal.
(c) One or more rows of $\mathbf{A}$ are zero.
(d) One or more columns of $\mathbf{A}$ are zero.

5.15. Show that all permutation matrices are of full rank. [Hint: Any permutation matrix can be expressed as a product of elementary matrices (of what type?).]

5.16. For any $m \times n$ matrix $\mathbf{A}$, what is the condition under which $\mathbf{A}\mathbf{A}^T$ is nonsingular? [Hint: Use Result 5.4(d).]

5.17. Use the matrices $\mathbf{A}$ and $\mathbf{B}$ in Exercise 5.1 to compute the determinant of:

(a) $\mathbf{A}^2$ (b) $\mathbf{A}^{10}$ (c) $\mathbf{A}\otimes\mathbf{A}$ (d) $\mathbf{A}\otimes\mathbf{B}$

5.18. Let $\mathbf{v}_1 = \begin{pmatrix} x_1 \\ y_1 \end{pmatrix}$, $\mathbf{v}_2 = \begin{pmatrix} x_2 \\ y_2 \end{pmatrix}$, $\mathbf{v}_3 = \begin{pmatrix} x_3 \\ y_3 \end{pmatrix}$, and $\mathbf{A} = \begin{pmatrix} 1 & 1 & 1 \\ x_1 & x_2 & x_3 \\ y_1 & y_2 & y_3 \end{pmatrix}$.

(a) Select some values for $\mathbf{v}_1$, $\mathbf{v}_2$, and $\mathbf{v}_3$, then draw a two-dimensional graph where each vector is represented by a point.
(b) Verify that the area of the triangle defined by the three points is equal to 0.5 $|\det(\mathbf{A})|$, which is one-half the absolute value of $\det(\mathbf{A})$.

5.19. For each of the vectors $\mathbf{u} = (2 \;\; -1 \;\; 0 \;\; 2)^T$ and $\mathbf{v} = (1 \;\; -1 \;\; 1 \;\; 1)^T$:

(a) Compute the L_1- norm, L_2-norm, and L_∞-norm.
(d) According to each of the above norms, which of the two vectors is larger than the other?

5.20. Verify Result 5.5(c)–(g) using $s = 2$ and the vectors in Exercise 5.19.

6 Matrix Inversion

In Chapter 2 we discussed matrix addition, subtraction, and multiplication. In this chapter we discuss matrix inversion, which is the matrix equivalent of scalar division. A matrix resulting from a matrix inversion is called a *matrix inverse*. Matrix inversion plays an important role in many applications. In Chapter 7, we give some examples illustrating the usefulness of matrix inverses, and we shall use matrix inverses repeatedly in the remaining chapters of the book.

Three types of matrix inverse (element-wise division, regular inverse, and generalized inverse) are presented in the following sections.

6.1. Element-Wise Division

In scalar algebra, the *reciprocal* or *inverse* of a nonzero number a is $1/a$. Can we define an inverse of a matrix $\mathbf{A}$ in an analogous way? For example, consider

$$\mathbf{A} = \begin{pmatrix} 5 & 2 & 1 & 4 \\ 2 & 3 & 6 & 5 \\ 7 & 6 & 3 & 2 \end{pmatrix},$$

which is the matrix given in (1.1), where a_{ij} is the cost of shipping one unit from Factory i to City j. Can we write $1/\mathbf{A} = (1/a_{ij})$, i.e.,

$$1/\mathbf{A} = \begin{pmatrix} 1/5 & 1/2 & 1/1 & 1/4 \\ 1/2 & 1/3 & 1/6 & 1/5 \\ 1/7 & 1/6 & 1/3 & 1/2 \end{pmatrix}?$$

Of course we can, and actually in this case the element $1/a_{ij}$ can be interpreted as the number of units that can be shipped from Factory i to City j for one dollar. We refer to this type of division as the *element-wise division*. Clearly, the element-wise division can be computed for any matrix as long as all of its elements are nonzero. This type of inverse, however, is not a common one. The most common definitions of matrix inverse are the regular inverse and generalized inverse, which are discussed in the following two sections.

6.2. The Regular Inverse

6.2.1. Definitions

In scalar algebra, we can always divide a scalar b by another scalar a as long as $a \neq 0$. We can write this as b/a or, equivalently $b \times a^{-1}$. The quantity a^{-1} is called the *reciprocal* or *inverse* of a. Thus, the inverse of a allows us to transform a division problem into a multiplication problem. Can we do the same in matrix algebra; that is, can we replace the scalars a and b by the matrices **A** and **B**? The answer is yes, but only if **A** has an inverse.

If a is a scalar, it is easy to define its inverse. It is the scalar whose product with a is 1. We write this as

$$a\, a^{-1} = a^{-1}\, a = 1.$$

To generalize this equation to matrices, we replace a by **A** and 1 by the identity matrix **I**, and write

$$\mathbf{A}\mathbf{A}^{-1} = \mathbf{A}^{-1}\mathbf{A} = \mathbf{I}. \tag{6.1}$$

The matrix $\mathbf{A}^{-1}$ (read as A-inverse) is the matrix whose product with **A** is **I**. This is the *regular inverse* of **A**, but we shall refer to it as the *inverse* of **A** for short.

As an example, let

$$\mathbf{A} = \begin{pmatrix} 1 & 1 \\ 3 & 4 \end{pmatrix}, \qquad \text{and} \qquad \mathbf{B} = \begin{pmatrix} 4 & -1 \\ -3 & 1 \end{pmatrix}, \tag{6.2}$$

and verify that

$$\mathbf{AB} = \mathbf{BA} = \begin{pmatrix} 1 & 0 \\ 0 & 1 \end{pmatrix} = \mathbf{I}_2.$$

We can then conclude that **B** is the inverse of **A** or, equivalently, that **A** is the inverse of **B**.

To further illustrate the relationship between **A** and its inverse $\mathbf{A}^{-1} = \mathbf{B}$, say, let us partition **A** in terms of its row vectors and **B** in terms of its column vectors, that is,

$$\mathbf{A} = \begin{pmatrix} \mathbf{a}_1^{\mathrm{T}} \\ \mathbf{a}_2^{\mathrm{T}} \\ \vdots \\ \mathbf{a}_n^{\mathrm{T}} \end{pmatrix} \qquad \text{and} \qquad \mathbf{B} = (\mathbf{b}_1 \quad \mathbf{b}_2 \quad \cdots \quad \mathbf{b}_n),$$

and compute the product

$$\mathbf{AB} = \begin{pmatrix} \mathbf{a}_1^{\mathrm{T}} \\ \mathbf{a}_2^{\mathrm{T}} \\ \vdots \\ \mathbf{a}_n^{\mathrm{T}} \end{pmatrix} (\mathbf{b}_1 \quad \mathbf{b}_2 \quad \cdots \quad \mathbf{b}_n) = \mathbf{I}$$

$$= \begin{pmatrix} \mathbf{a}_1^{\mathrm{T}}\mathbf{b}_1 & \mathbf{a}_1^{\mathrm{T}}\mathbf{b}_2 & \cdots & \mathbf{a}_1^{\mathrm{T}}\mathbf{b}_n \\ \mathbf{a}_2^{\mathrm{T}}\mathbf{b}_1 & \mathbf{a}_2^{\mathrm{T}}\mathbf{b}_2 & \cdots & \mathbf{a}_2^{\mathrm{T}}\mathbf{b}_n \\ \vdots & \vdots & \ddots & \vdots \\ \mathbf{a}_n^{\mathrm{T}}\mathbf{b}_1 & \mathbf{a}_n^{\mathrm{T}}\mathbf{b}_2 & \cdots & \mathbf{a}_n^{\mathrm{T}}\mathbf{b}_n \end{pmatrix} = \begin{pmatrix} 1 & 0 & \cdots & 0 \\ 0 & 1 & \cdots & 0 \\ \vdots & \vdots & \ddots & \vdots \\ 0 & 0 & \cdots & 1 \end{pmatrix}.$$

It follows then that the inner product of the ith row of **A** and the jth column of **B** is 1 when $i = j$ and 0 when $i \neq j$. Therefore, when $i \neq j$, the ith row of **A** and the jth column of **B** are orthogonal to each other. The same relationships also hold for the columns of **A** and the rows of **B** (why?). Furthermore, the rows (columns) of **A** can be thought of as a basis for the vector space generated by the rows (columns) of **A**, and the rows (columns) of **B** can be thought of as another basis for the same vector space.

In the remainder of this section we address three important questions:

1. Does the inverse always exist?
2. When it exists, how is $\mathbf{A}^{-1}$ computed?
3. What are the properties of the inverse when it exists?

6.2.2. Existence of Matrix Inverse

Matrix inverses do not always exist. First, we see from (6.1) that, for $\mathbf{A}^{-1}$ to exist, **A** and $\mathbf{A}^{-1}$ must be conformable for multiplication, which implies that **A** must be a square matrix. Thus, the regular inverse is defined only for square matrices. If **A** is not square, a different inverse, called a *generalized inverse*, can be defined. Generalized inverses are dealt with in Section 6.3.

Second, we know from scalar algebra that a^{-1} exists if and only if $a \neq 0$. The analogous condition for a square matrix is given in the following result.

Result 6.1. For any $n \times n$ matrix **A**, $\mathbf{A}^{-1}$ exists if and only if $\det(\mathbf{A}) \neq 0$, or, equivalently, if and only if $\mathrm{r}(\mathbf{A}) = n$.

Thus, the determinant of a matrix determines whether or not its inverse exists. Recall from Section 5.2 that a matrix whose determinant is 0 is called a singular matrix. Therefore, Result 6.1 provides another equivalent definition of singular and nonsingular matrices, namely, a singular matrix is a matrix that has no regular inverse, and a nonsingular matrix is a matrix that has a regular inverse. For example, the matrices **A** and **B** in (6.2) are nonsingular because they have nonzero determinants, but

$$\mathbf{C} = \begin{pmatrix} 2 & 1 \\ 4 & 2 \end{pmatrix}, \tag{6.3}$$

for example, is singular because det($\mathbf{C}$) = 0 or because r($\mathbf{C}$) = 1; hence $\mathbf{C}^{-1}$ does not exist.

6.2.3. An Algorithm for Computing Matrix Inverse

There are several algorithms for computing the inverse of a matrix if it exists. These algorithms involve a great deal of calculation, so computing matrix inverses is usually done by computer (especially for large matrices). In the appendix to this chapter we present a simple algorithm for computing the inverse of any square matrix, when it exists, and illustrate it by a few examples. Here, however, we give a formula for computing the inverse of 2×2 matrices. Suppose that

$$\mathbf{A} = \begin{pmatrix} a_{11} & a_{12} \\ a_{21} & a_{22} \end{pmatrix};$$

then its inverse is given by

$$\mathbf{A}^{-1} = \frac{1}{\det(\mathbf{A})} \begin{pmatrix} a_{22} & -a_{12} \\ -a_{21} & a_{11} \end{pmatrix}, \tag{6.4}$$

where det($\mathbf{A}$) = $a_{11}\, a_{22} - a_{21}\, a_{12}$, as defined in (5.2). The reader can verify that $\mathbf{A}^{-1}\mathbf{A} = \mathbf{A}\mathbf{A}^{-1} = \mathbf{I}$. You can see from (6.4) why the inverse does not exist when the determinant is 0. If you apply this formula to the matrix $\mathbf{A}$ in (6.2), you will obtain the matrix $\mathbf{B}$ (and the reverse operation is also true). On the other hand, we cannot apply (6.4) to the matrix $\mathbf{C}$ in (6.3), because det($\mathbf{C}$) = 0 and we cannot divide by a zero determinant.

6.2.4. Properties of Matrix Inverse

Matrix inversion has several useful and interesting properties, some of which are given in the following series of results.

Result 6.2. The matrix inverse, provided it exists, has the following properties:
(a) The inverse is unique.
(b) If $\mathbf{D} = diag(d_{ii})$, then $\mathbf{D}^{-1} = diag(1/d_{ii})$, that is, the inverse of a diagonal matrix $\mathbf{D}$ is a diagonal matrix containing the reciprocals of the corresponding diagonal elements of $\mathbf{D}$.
(c) $\det(\mathbf{A}^{-1}) = [\det(\mathbf{A})]^{-1}$; that is, the determinant of the inverse is the inverse of the determinant.
(d) The reflexive property: $(\mathbf{A}^{-1})^{-1} = \mathbf{A}$; that is, the inverse of $\mathbf{A}^{-1}$ is $\mathbf{A}$.

(e) The reversal property: $(\mathbf{AB})^{-1} = \mathbf{B}^{-1}\mathbf{A}^{-1}$, provided that **A** and **B** are square matrices. In other words, the inverse of a product is the product of the inverses but in reverse order.
(f) $(\mathbf{A}^{\mathrm{T}})^{-1} = (\mathbf{A}^{-1})^{\mathrm{T}}$; that is, the inverse of the transpose is the transpose of the inverse.
(g) If **A** is an orthogonal matrix, then $\mathbf{A}^{-1} = \mathbf{A}^{\mathrm{T}}$.
(h) $(\mathbf{A}\otimes\mathbf{B})^{-1} = \mathbf{A}^{-1}\otimes\mathbf{B}^{-1}$, provided that **A** and **B** are square matrices; that is, the inverse of a Kronecker product is the product of the inverses, but the order of matrices is not reversed.

Result 6.3. All elementary matrices and their products are nonsingular.[1]

Result 6.4. The inverse of a partitioned matrix: For any partitioned matrix

$$\mathbf{P} = \begin{pmatrix} \mathbf{A} & \mathbf{B} \\ \mathbf{C} & \mathbf{D} \end{pmatrix}, \tag{6.5}$$

where **P**, **A**, and **D** are square matrices, we have:
(a) Provided that **P** and **A** are nonsingular,

$$\mathbf{P}^{-1} = \begin{pmatrix} \mathbf{E} & \mathbf{F} \\ \mathbf{G} & \mathbf{H} \end{pmatrix}, \tag{6.6}$$

where

$$\mathbf{H} = (\mathbf{D} - \mathbf{C}\mathbf{A}^{-1}\mathbf{B})^{-1}, \qquad \mathbf{G} = -\mathbf{H}\mathbf{C}\mathbf{A}^{-1},$$
$$\mathbf{F} = -\mathbf{A}^{-1}\mathbf{B}\mathbf{H}, \qquad \mathbf{E} = \mathbf{A}^{-1} - \mathbf{F}\mathbf{C}\mathbf{A}^{-1}.$$

(b) Provided that **P** and **D** are nonsingular,

$$\mathbf{P}^{-1} = \begin{pmatrix} \mathbf{E} & \mathbf{F} \\ \mathbf{G} & \mathbf{H} \end{pmatrix}, \tag{6.7}$$

where

$$\mathbf{E} = (\mathbf{A} - \mathbf{B}\mathbf{D}^{-1}\mathbf{C})^{-1}, \qquad \mathbf{F} = -\mathbf{E}\mathbf{B}\mathbf{D}^{-1},$$
$$\mathbf{G} = -\mathbf{D}^{-1}\mathbf{C}\mathbf{E}, \qquad \mathbf{H} = \mathbf{D}^{-1} - \mathbf{G}\mathbf{B}\mathbf{D}^{-1}.$$

The above formulas are very useful, for example, in simplifying many expressions in regression and multivariate analysis. When the matrices **B** and **C** in (6.5) are vectors **b** and $\mathbf{c}^{\mathrm{T}}$ (hence **D** is a scalar d), the above formulae simplify considerably. In this case, (6.5) and (6.6) become

---

[1] Elementary matrices are defined in Section 3.6.

$$\mathbf{P} = \begin{pmatrix} \mathbf{A} & \mathbf{b} \\ \mathbf{c}^{\mathrm{T}} & d \end{pmatrix} \tag{6.8}$$

and

$$\mathbf{P}^{-1} = \begin{pmatrix} \mathbf{E} & \mathbf{f} \\ \mathbf{g}^{\mathrm{T}} & h \end{pmatrix}, \tag{6.9}$$

where

$$h = (d - \mathbf{c}^{\mathrm{T}}\mathbf{A}^{-1}\mathbf{b})^{-1}, \qquad \mathbf{g}^{\mathrm{T}} = -h\mathbf{c}^{\mathrm{T}}\mathbf{A}^{-1},$$
$$\mathbf{f} = -\mathbf{A}^{-1}\mathbf{b}h, \qquad \mathbf{E} = \mathbf{A}^{-1} - \mathbf{f}\mathbf{c}^{\mathrm{T}}\mathbf{A}^{-1}.$$

Also, (6.7) becomes

$$\mathbf{P}^{-1} = \begin{pmatrix} \mathbf{E} & \mathbf{f} \\ \mathbf{g}^{\mathrm{T}} & h \end{pmatrix}, \tag{6.10}$$

where

$$\mathbf{E} = (\mathbf{A} - d^{-1}\mathbf{b}\mathbf{c}^{\mathrm{T}})^{-1}, \qquad \mathbf{f} = -d^{-1}\mathbf{E}\mathbf{b},$$
$$\mathbf{g}^{\mathrm{T}} = -d^{-1}\mathbf{c}^{\mathrm{T}}\mathbf{E}, \qquad h = (1 - \mathbf{g}^{\mathrm{T}}\mathbf{b})/d.$$

Equations (6.9) and (6.10) are useful from a computational point of view. For example, suppose we have already computed the inverse of the submatrix $\mathbf{A}$ in (6.8) and we now wish to compute the inverse of $\mathbf{P}$. We may achieve computational simplifications if we use (6.9) instead of computing $\mathbf{P}^{-1}$ from scratch. Thus, (6.9) can be considered as an *updating formula* because it can be used to update the inverse of a matrix using the inverse of a submatrix. Another very useful updating formula is given in the following result.

Result 6.5. Let $\mathbf{A}$ and $\mathbf{D}$ be nonsingular matrices of order $p \times p$ and $q \times q$, respectively, and $\mathbf{B}$ and $\mathbf{C}$ be matrices of order $p \times q$. Then, provided that the inverses exist,

$$(\mathbf{A} + \mathbf{B}\mathbf{D}\mathbf{C}^{\mathrm{T}})^{-1} = \mathbf{A}^{-1} - \mathbf{A}^{-1}\mathbf{B}(\mathbf{D}^{-1} + \mathbf{C}^{\mathrm{T}}\mathbf{A}^{-1}\mathbf{B})^{-1}\mathbf{C}^{\mathrm{T}}\mathbf{A}^{-1}. \tag{6.11}$$

Formula (6.11) is particularly useful when $\mathbf{B}$, $\mathbf{D}$, and $\mathbf{C}$ are of simple forms. For example, when $\mathbf{B}$ and $\mathbf{C}$ are replaced by column vectors $\mathbf{b}$ and $\mathbf{c}$, respectively, and $\mathbf{D}$ is replaced by the scalar 1, (6.11) reduces to

$$(\mathbf{A} + \mathbf{b}\mathbf{c}^{\mathrm{T}})^{-1} = \mathbf{A}^{-1} - \frac{\mathbf{A}^{-1}\mathbf{b}\mathbf{c}^{\mathrm{T}}\mathbf{A}^{-1}}{1 + \mathbf{c}^{\mathrm{T}}\mathbf{A}^{-1}\mathbf{b}}. \tag{6.12}$$

Applications of the above updating formulas are given in Section 8.4.

Now that matrix inverse is defined, we are able to discuss other properties of the determinant that involve the matrix inverse.

Result 6.6. Let $\mathbf{A}$ be a nonsingular matrix of order $n \times n$ and $\mathbf{b}$ be a vector of order $n \times 1$. Then

$$\det(\mathbf{A} + \mathbf{b}\mathbf{b}^T) = \det(\mathbf{A}) \det(1 + \mathbf{b}^T\mathbf{A}^{-1}\mathbf{b}).$$

Result 6.7. The determinant of a partitioned matrix: For any partitioned matrix of the form

$$\mathbf{P} = \begin{pmatrix} \mathbf{A} & \mathbf{B} \\ \mathbf{C} & \mathbf{D} \end{pmatrix},$$

where $\mathbf{P}$, $\mathbf{A}$, and $\mathbf{D}$ are square matrices, we have:
(a) If $\mathbf{A}$ is nonsingular, then

$$\det(\mathbf{P}) = \det(\mathbf{A}) \det(\mathbf{D} - \mathbf{C}\mathbf{A}^{-1}\mathbf{B}).$$

(b) If $\mathbf{D}$ is nonsingular, then

$$\det(\mathbf{P}) = \det(\mathbf{D}) \det(\mathbf{A} - \mathbf{B}\mathbf{D}^{-1}\mathbf{C}).$$

By this result we can express the determinant of a large matrix in terms of the determinants of smaller matrices.

*6.3. Generalized Inverse

As mentioned in the previous section, the regular inverse is defined only for square nonsingular matrices. In this section, we generalize the definition of the regular inverse to rectangular and singular matrices. Our discussion here is both elementary and brief. For a more comprehensive account of generalized inverse (or g-inverse, for short) the reader is referred to texts such as Rao and Mitra (1971), Albert (1972), and Searle (1982).

We start by a definition of generalized inverse and, in keeping with our treatment of the regular inverse, we address the following questions:

1. Does a g-inverse always exist?
2. Is it unique?
3. How it is computed?
4. What are the properties of a g-inverse?

We then present some particular types of g-inverse and conclude with a discussion of how a g-inverse can be used to obtain a result known as the full-rank factorization result.

6.3.1. Definitions

Let $\mathbf{A}$ be an $n \times m$ matrix; then if a matrix $\mathbf{G}$ satisfies

$$\mathbf{AGA} = \mathbf{A}, \tag{6.13}$$

$\mathbf{G}$ is called a *generalized inverse* or *g-inverse* of $\mathbf{A}$.

To distinguish the regular inverse from a g-inverse we write $\mathbf{G} = \mathbf{A}^-$ (pronounced as A-minus). Thus, $\mathbf{A}^-$ denotes a g-inverse of $\mathbf{A}$ and (6.13) becomes

$$\mathbf{AA}^-\mathbf{A} = \mathbf{A}. \tag{6.14}$$

It is implicit in (6.14) that $\mathbf{A}^-$ is of order $m \times n$ (the reverse of the order of $\mathbf{A}$). Note also that when $\mathbf{A}$ is square and nonsingular, $\mathbf{A}^{-1}$ exists and, by substituting $\mathbf{A}^{-1}$ for $\mathbf{A}^-$ in (6.14), we see that $\mathbf{A}^{-1}$ satisfies (6.14) because $\mathbf{AA}^{-1}\mathbf{A} = \mathbf{IA} = \mathbf{A}$. Additionally, $\mathbf{A}^-\mathbf{A} \neq \mathbf{I}$ unless $\mathbf{A}$ is nonsingular, in which case $\mathbf{A}^- = \mathbf{A}^{-1}$. Therefore, the regular inverse is a special case of the g-inverse, or, in other words, (6.14) is indeed a generalization of the regular inverse.

6.3.2. Existence and Uniqueness of G-Inverse

Let $\mathbf{A}$ be an $n \times m$ matrix of which we wish to obtain a g-inverse. Suppose that $\mathrm{r}(\mathbf{A}) = k > 0$. Then, by Result 3.5(g), there are k linearly independent rows and k linearly independent columns of $\mathbf{A}$. Suppose, for a moment, that these are the first k rows and k columns of $\mathbf{A}$. In this case, $\mathbf{A}$ can be partitioned as

$$\mathbf{A} = \begin{pmatrix} \mathbf{A}_{11} & \mathbf{A}_{12} \\ \mathbf{A}_{21} & \mathbf{A}_{22} \end{pmatrix}, \tag{6.15}$$

where the submatrix $\mathbf{A}_{11}$ is of order $k \times k$ and of rank k. Since $\mathbf{A}$ is of order $n \times m$ it follows that $\mathbf{A}_{12}$ is of order $k \times (m - k)$, $\mathbf{A}_{21}$ is of order $(n - k) \times k$, and $\mathbf{A}_{22}$ is of order $(n - k) \times (m - k)$. Since $\mathrm{r}(\mathbf{A}_{11}) = k$, then by Result 6.1, $\mathbf{A}_{11}$ is nonsingular, so its (regular) inverse $\mathbf{A}_{11}^{-1}$ exists. We now have the following result.

Result 6.8. Let $\mathbf{A}$ be a matrix of rank $k > 0$. If $\mathbf{A}$ is partitioned as in (6.15), then:
(a) A g-inverse of $\mathbf{A}$ always exists and is given by

$$\mathbf{G} = \begin{pmatrix} \mathbf{G}_{11} & \mathbf{G}_{12} \\ \mathbf{G}_{21} & \mathbf{G}_{22} \end{pmatrix} = \mathbf{A}^-, \tag{6.16}$$

where $\mathbf{G}_{12}$, $\mathbf{G}_{21}$, and $\mathbf{G}_{22}$ are arbitrary matrices of appropriate order and

$$\mathbf{G}_{11} = \mathbf{A}_{11}^{-1} - \mathbf{A}_{11}^{-1}\mathbf{A}_{12}\mathbf{G}_{21} - \mathbf{G}_{12}\mathbf{A}_{21}\mathbf{A}_{11}^{-1} - \mathbf{A}_{11}^{-1}\mathbf{A}_{12}\mathbf{G}_{22}\mathbf{A}_{21}\mathbf{A}_{11}^{-1}. \tag{6.17}$$

(b) $\mathbf{A}^-$ is not unique.
(c) A simple g-inverse of $\mathbf{A}$ is given by

$$\mathbf{A}^- = \begin{pmatrix} \mathbf{A}_{11}^{-1} & \mathbf{0} \\ \mathbf{0} & \mathbf{0} \end{pmatrix}. \tag{6.18}$$

Thus, unlike the regular inverse, which is unique if it exists, a g-inverse always exists for matrices of nonzero rank, but is not unique. The arbitrariness of $\mathbf{G}_{12}$, $\mathbf{G}_{21}$, and $\mathbf{G}_{22}$ gives rise to the last two parts of the result. Part (c) follows by setting the arbitrary matrices to zero, yielding $\mathbf{G}_{11} = \mathbf{A}_{11}^{-1}$; then (6.16) reduces to (6.18). The reader can verify that $\mathbf{A}^-$ in (6.18) is indeed a g-inverse of $\mathbf{A}$ in (6.15) by showing that $\mathbf{A}$ and $\mathbf{A}^-$ satisfy (6.14).

It is time for a numerical example. We have seen that the matrix

$$\mathbf{C} = \begin{pmatrix} 2 & 1 \\ 4 & 2 \end{pmatrix},$$

in (6.3), does not have a regular inverse, so let us find a g-inverse of $\mathbf{C}$. We first need to partition $\mathbf{C}$ as in (6.15). We have

$$\begin{aligned} \mathbf{C} &= \begin{pmatrix} \mathbf{C}_{11} & \mathbf{C}_{12} \\ \mathbf{C}_{21} & \mathbf{C}_{22} \end{pmatrix} \\ &= \begin{pmatrix} 2 & 1 \\ 4 & 2 \end{pmatrix}. \end{aligned}$$

The rank of $\mathbf{C}$ is $k = 1$. In the above partition of $\mathbf{C}$, the submatrix $\mathbf{C}_{11}$ (consisting of the elements in the first k rows and the first k columns of $\mathbf{C}$) is, in this case, just the scalar 2, and all the other submatrices are also scalars. Thus, $\mathbf{C}_{11}^{-1} = 0.5$ and, using (6.18), we have

$$\mathbf{C}^- = \begin{pmatrix} 0.5 & 0 \\ 0 & 0 \end{pmatrix}, \tag{6.19}$$

which is a g-inverse of $\mathbf{C}$. To obtain another g-inverse of $\mathbf{C}$, we can use arbitrary values for the submatrices $\mathbf{G}_{21}$, $\mathbf{G}_{12}$, and $\mathbf{G}_{22}$ in (6.17). For example, we can set $\mathbf{G}_{21} = \mathbf{G}_{12} = \mathbf{G}_{22} = 2$, which gives us $\mathbf{G}_{11} = -6.5$. Then (6.16) becomes

$$\mathbf{C}^- = \begin{pmatrix} -6.5 & 2.0 \\ 2.0 & 2.0 \end{pmatrix}, \tag{6.20}$$

which is another g-inverse of $\mathbf{C}$. The reader can verify that the two g-inverses in (6.19) and (6.20) satisfy $\mathbf{C}\mathbf{C}^-\mathbf{C} = \mathbf{C}$.

6.3.3. An Algorithm for Computing a G-Inverse

There are several algorithms for computing a g-inverse. A g-inverse also requires more computations than the regular inverse. Here, we present a simple (but not necessarily computationally efficient) algorithm for computing a g-inverse. This algorithm is developed as follows.

Result 6.8 gives a simple algorithm for computing a g-inverse, but it requires the square matrix in the intersection of the first k rows and k columns of $\mathbf{A}$ to be nonsingular so that $\mathbf{A}$ can be partitioned as in (6.15). Suppose now that a matrix $\mathbf{B}$, of which we wish to compute a g-inverse, does not satisfy this requirement. We know from Result 3.5(g) that there are k linearly independent rows and k linearly independent columns of $\mathbf{B}$. We can then permute the rows and columns of $\mathbf{B}$ so that the first k rows and k columns of the permuted matrix can be partitioned in a form similar to that in (6.15). The permutation matrices $\mathbf{R}$ and $\mathbf{C}$ that are needed to perform this task can be found as in (3.13) and (3.14). Therefore, the matrix

$$\mathbf{RBC} = \mathbf{A} \tag{6.21}$$

can be partitioned in the form given by (6.15). We can now use Result 6.8 to compute $\mathbf{A}^-$, a g-inverse of $\mathbf{A}$. Finally, a g-inverse of $\mathbf{B}$ can be computed from a g-inverse of $\mathbf{A}$ as stated in the following result.

Result 6.9. Let $\mathbf{B}$ be an $n \times m$ matrix of rank $k > 0$; then

(a) The matrix $\mathbf{B}$ can be written as

$$\mathbf{B} = \mathbf{R}^{-1}\mathbf{A}\mathbf{C}^{-1}, \tag{6.22}$$

where $\mathbf{R}$ and $\mathbf{C}$ are the permutation matrices required for $\mathbf{A}$ to be partitioned as in (6.15).

(b) A g-inverse of $\mathbf{B}$ is given by

$$\mathbf{B}^- = \mathbf{C}\mathbf{A}^-\mathbf{R}, \tag{6.23}$$

where $\mathbf{A}^-$ is a g-inverse of $\mathbf{A}$ in (6.22).

Note that since a permutation matrix is a product of elementary matrices, then $\mathbf{R}$ and $\mathbf{C}$ are nonsingular and their regular inverses exist by Result 6.3. Part (a) follows immediately after a pre- and post-multiplication of both sides of (6.21) by $\mathbf{R}^{-1}$ and $\mathbf{C}^{-1}$, respectively. Part (b) follows because

$$\begin{aligned}\mathbf{BB}^-\mathbf{B} &= (\mathbf{R}^{-1}\mathbf{A}\mathbf{C}^{-1})(\mathbf{C}\mathbf{A}^-\mathbf{R})(\mathbf{R}^{-1}\mathbf{A}\mathbf{C}^{-1})\\ &= \mathbf{R}^{-1}(\mathbf{A}\mathbf{A}^-\mathbf{A})\mathbf{C}^{-1}\\ &= \mathbf{R}^{-1}\mathbf{A}\mathbf{C}^{-1} = \mathbf{B}.\end{aligned}$$

6.3.4. Properties of G-Inverse

A g-inverse has the following properties.

Result 6.10. A g-inverse, $\mathbf{A}^-$, of an $n \times m$ matrix $\mathbf{A}$ of rank k is of order $m \times n$, and satisfies:

(a) $\mathrm{r}(\mathbf{A}) \le \mathrm{r}(\mathbf{A}^-) \le \min(n, m)$.
(b) If $\mathbf{A}$ is symmetric, $\mathbf{A}^-$ is not necessarily symmetric. However, a symmetric g-inverse can always be obtained.
(c) If $\mathbf{A}$ is symmetric, then both $\mathbf{A}^-$ and $(\mathbf{A}^-)^\mathrm{T}$ are g-inverses of $\mathbf{A}$.
(d) The square matrices $\mathbf{T} = \mathbf{A}^-\mathbf{A}$ and $\mathbf{U} = \mathbf{A}\mathbf{A}^-$ are idempotent and have rank k.
(e) The square matrices $(\mathbf{I} - \mathbf{T})$ and $(\mathbf{I} - \mathbf{U})$ are idempotent with $\mathrm{r}(\mathbf{I} - \mathbf{T}) = m - k$ and $\mathrm{r}(\mathbf{I} - \mathbf{U}) = n - k$.
(f) A g-inverse of an idempotent matrix is the matrix itself.

Since $\mathbf{A}_{11}$ in (6.18) has rank k, $\mathrm{r}(\mathbf{A}^-)$ cannot be less than k. Although $\mathbf{A}$ may be rank-deficient, by suitable choices for the arbitrary matrices $\mathbf{G}_{12}$, $\mathbf{G}_{21}$, and $\mathbf{G}_{22}$ in (6.16) we can always construct a g-inverse of $\mathbf{A}$ which is either full-column or full-row rank and Part (a) follows. As stated in Part (b), not all g-inverses of a symmetric matrix are symmetric. To obtain a symmetric g-inverse, as claimed in Part (b), we apply the same permutation to the rows and columns. The resultant $\mathbf{A}_{11}$ will then be symmetric, which will give rise to a symmetric g-inverse.

Part (c) is self-explanatory. For Part (d), recall from Section 2.3.2 that an idempotent matrix $\mathbf{T}$ must satisfy $\mathbf{TT} = \mathbf{T}$. Thus, we have $\mathbf{TT} = \mathbf{A}^-(\mathbf{A}\mathbf{A}^-\mathbf{A}) = \mathbf{A}^-\mathbf{A} = \mathbf{T}$. Combining Result 5.1(e) and Result 6.10(d), we have $\mathrm{r}(\mathbf{T}) = \mathrm{tr}(\mathbf{T}) = k$. Note that $\mathbf{T}$ and $\mathbf{U}$ are not of the same order unless $n = m$. A similar argument applies to $\mathbf{U}$, $(\mathbf{I} - \mathbf{T})$, and $(\mathbf{I} - \mathbf{U})$. Finally, for Part (f), since $\mathbf{T}$ is idempotent, it follows that $\mathbf{TTT} = \mathbf{T}$, which means that $\mathbf{T}$ is a g-inverse of itself.

6.3.5. Particular Types of G-Inverse

Our definition of g-inverse so far requires only that $\mathbf{A}^-$ satisfies (6.14), that is,

$$\mathbf{A}\mathbf{A}^-\mathbf{A} = \mathbf{A}. \tag{6.24}$$

Additional conditions, however, can be imposed on $\mathbf{A}^-$. These additional conditions define several other types of g-inverse. For example, if in addition to (6.24) the matrix $\mathbf{A}^-$ also satisfies

$$\mathbf{A}^-\mathbf{A}\mathbf{A}^- = \mathbf{A}, \tag{6.25}$$

the inverse is called a *reflexive* g-inverse. A reflexive g-inverse that also satisfies

$$(\mathbf{A}\mathbf{A}^-)^\mathrm{T} = \mathbf{A}\mathbf{A}^- \tag{6.26}$$

and

$$(\mathbf{A}^-\mathbf{A})^\mathrm{T} = \mathbf{A}^-\mathbf{A} \tag{6.27}$$

is called the *Moore-Penrose inverse* and is usually denoted by $\mathbf{A}^+$. Thus, $\mathbf{A}^+$ must satisfy all four conditions (6.24)–(6.27). As a result, it has the following properties.

Result 6.11. Let $\mathbf{A}^+$ be the Moore-Penrose inverse of an $n \times m$ matrix $\mathbf{A}$. Then
(a) $\mathbf{A}^+$ is a unique matrix of order $m \times n$.
(b) If $\mathbf{A}$ is of full-column rank, then $\mathbf{A}^+ = (\mathbf{A}^T\mathbf{A})^{-1}\mathbf{A}^T$, hence $\mathbf{A}^+\mathbf{A} = \mathbf{I}_m$.
(c) If $\mathbf{A}$ is of full-row rank, then $\mathbf{A}^+ = \mathbf{A}^T(\mathbf{A}\mathbf{A}^T)^{-1}$, hence $\mathbf{A}\mathbf{A}^+ = \mathbf{I}_n$.

Part (a) asserts that the Moore-Penrose inverse is unique (like the regular inverse but unlike any of the other g-inverses mentioned above). Parts (b) and (c) indicate that the Moore-Penrose inverse can be computed in a straightforward manner in cases where $\mathbf{A}$ is of either full-column or full-row rank. Note that if $\mathbf{A}$ is of full-column rank, then, by Result 5.4(d), the $m \times m$ matrix $\mathbf{A}^T\mathbf{A}$ is nonsingular; hence its regular inverse exists. Similarly, if $\mathbf{A}$ is of full-row rank, then the $n \times n$ matrix $\mathbf{A}\mathbf{A}^T$ is nonsingular; hence its regular inverse exists.

6.3.6. Full-Rank Factorization

We now return to the rank factorization in Result 3.5(h), as we promised at the end of Section 3.4, and make a remark about how the matrices $\mathbf{P}$ and $\mathbf{Q}$ are found. As we mentioned above, the first step in computing a g-inverse of any matrix $\mathbf{B}$ is usually to permute the rows and/or columns of $\mathbf{B}$, if necessary, to obtain a matrix $\mathbf{A}$ that can be partitioned in the form given by (6.15). Thus, by (6.15) and (6.21), $\mathbf{A}$ can be expressed as

$$\mathbf{A} = \mathbf{RBC} = \begin{pmatrix} \mathbf{A}_{11} & \mathbf{A}_{12} \\ \mathbf{A}_{21} & \mathbf{A}_{22} \end{pmatrix}. \tag{6.28}$$

Now, since $\mathbf{A}_{11}$ is a $k \times k$ matrix of rank k, it follows that the last $n - k$ rows of $\mathbf{A}$ are linearly dependent on the first k rows. Similarly, the last $m - k$ columns are dependent on the first k columns. Therefore, if we apply the row and column operations on $\mathbf{A}$ that are needed to replace $\mathbf{A}_{11}$ by an identity matrix of order k, these operations will replace the matrices $\mathbf{A}_{12}$, $\mathbf{A}_{21}$, and $\mathbf{A}_{22}$ by zero matrices of respective order. Let $\mathbf{L}$ be the product of the row-elementary matrices and $\mathbf{M}$ be the product of the column-elementary matrices required to replace $\mathbf{A}_{11}$ by $\mathbf{I}_k$. By Result 6.3, both $\mathbf{L}$ and $\mathbf{M}$ are nonsingular. Then, by pre- and post-multiplying (6.28) by $\mathbf{L}$ and $\mathbf{M}$, respectively, we obtain

$$\mathbf{LAM} = \mathbf{LRBCM} = \begin{pmatrix} \mathbf{I}_k & \mathbf{0} \\ \mathbf{0} & \mathbf{0} \end{pmatrix},$$

from which it follows that

$$\mathbf{B} = \mathbf{R}^{-1}\mathbf{L}^{-1}\begin{pmatrix} \mathbf{I}_k & \mathbf{0} \\ \mathbf{0} & \mathbf{0} \end{pmatrix}\mathbf{M}^{-1}\mathbf{C}^{-1} = \mathbf{P}\begin{pmatrix} \mathbf{I}_k & \mathbf{0} \\ \mathbf{0} & \mathbf{0} \end{pmatrix}\mathbf{Q}, \tag{6.29}$$

where $\mathbf{P} = \mathbf{R}^{-1}\mathbf{L}^{-1}$ and $\mathbf{Q} = \mathbf{M}^{-1}\mathbf{C}^{-1}$. Equation (6.29) is the same as Result 3.5(h), and this is one way to compute the matrices **P** and **Q**.

Result 3.5(h) is useful in at least two respects. First, if we partition **P** and **Q** as

$$\mathbf{P} = (\mathbf{P}_1 \ : \ \mathbf{P}_2) \quad \text{and} \quad \mathbf{Q} = \begin{pmatrix} \mathbf{Q}_1 \\ \mathbf{Q}_2 \end{pmatrix}, \tag{6.30}$$

where $\mathbf{P}_1$ is of order $n \times k$ and $\mathbf{Q}_1$ is of order $k \times m$, and then carry out the matrix multiplications in (6.29), we obtain

$$\mathbf{B} = \mathbf{P}_1 \, \mathbf{Q}_1, \tag{6.31}$$

which indicates that any matrix **B** of rank k can be factored into the product of two matrices. The factorization in (6.31) is known as the *full-rank factorization.*

Second, Result 3.5(h) is also useful in the computation of a g-inverse. Let

$$\mathbf{D} = \begin{pmatrix} \mathbf{I}_k & \mathbf{0} \\ \mathbf{0} & \mathbf{0} \end{pmatrix}, \tag{6.32}$$

so that (6.29) becomes $\mathbf{B} = \mathbf{PDQ}$ or, equivalently, $\mathbf{D} = \mathbf{P}^{-1}\mathbf{B}\mathbf{Q}^{-1}$. The following result can be easily derived.

Result 6.12. Let **B** be an $n \times m$ matrix of rank k written as in (6.31), **D** be defined as in (6.32), and **C** and **R** be permutation matrices that transform **B** into a matrix that can be suitably partitioned (see (6.21)). Then

(a) $\mathbf{D}^{\mathrm{T}}$ is a g-inverse of **D**.

(b) $\mathbf{Q}^{-1}\mathbf{D}^{\mathrm{T}}\mathbf{P}^{-1} = \mathbf{CMD}^{\mathrm{T}}\mathbf{LR}$ is a g-inverse of **B**.

Appendix: Computing Matrix Inverse

This appendix presents a simple algorithm for computing matrix inverses and illustrates it by examples. Let us assume for the moment that the inverse $\mathbf{A}^{-1}$ of an $n \times n$ matrix **A** exists. To compute $\mathbf{A}^{-1}$, one may execute the following steps:

Step 1. Create a new matrix by appending to the matrix **A** an identity matrix of the same order. The new matrix $(\mathbf{A} : \mathbf{I})$ is called the *augmented* matrix.

Step 2. Systematically apply to the augmented matrix a sequence of row operations[2] that would reduce the matrix **A** to **I**. The same sequence of row operations will then replace **I** in the augmented matrix by $\mathbf{A}^{-1}$.

This algorithm can be used to compute the inverse of a square matrix of any order if the inverse exists. We illustrate it by two examples; in the first example $\mathbf{A}^{-1}$ exists and in the second it does not.

---

2 See the appendix to Chapter 3.

When the inverse exists. Suppose we wish to compute the inverse of

$$\mathbf{A} = \begin{pmatrix} 2 & 6 & 4 \\ 5 & 30 & 15 \\ 6 & 3 & 9 \end{pmatrix}. \tag{6.33}$$

We first need to create the augmented matrix

$$\mathbf{A}_1 = (\mathbf{A} \;:\; \mathbf{I}) = \begin{pmatrix} 2 & 6 & 4 & : & 1 & 0 & 0 \\ 5 & 30 & 15 & : & 0 & 1 & 0 \\ 6 & 3 & 9 & : & 0 & 0 & 1 \end{pmatrix}. \tag{6.34}$$

We then apply to $\mathbf{A}_1$ a sequence of row operations that reduces $\mathbf{A}$ to $\mathbf{I}$. We do this in a systematic way, taking the columns of $\mathbf{A}$ one at a time and reducing them to the corresponding columns of the identity matrix.

Let us start with the first column. The elements 2, 5, and 6 must be replaced by 1, 0, and 0, respectively. To replace the a_{11} element by 1, we multiply the first row of $\mathbf{A}_1$ by 0.5. To replace the a_{21} element by 0, we multiply the first row of $\mathbf{A}_1$ by −2.5 and add the result to row 2. To replace the a_{31} element by 0, we multiply the first row of $\mathbf{A}_1$ by −3 and add the result to row 3. Using the shorthand notation introduced in the appendix to Chapter 3, we obtain

$$\mathbf{A}_2 = \begin{pmatrix} 1 & 3 & 2 & : & 0.5 & 0 & 0 \\ 0 & 15 & 5 & : & -2.5 & 1 & 0 \\ 0 & -15 & -3 & : & -3.0 & 0 & 1 \end{pmatrix} \begin{matrix} 0.5\mathbf{r}_1 \\ \mathbf{r}_2 - 2.5\mathbf{r}_1 \\ \mathbf{r}_3 - 3\mathbf{r}_1 \end{matrix}. \tag{6.35}$$

We now work on the second column. To replace the first element by 0, we multiply the second row of $\mathbf{A}_2$ by −3/15 and add the result to row 1. Multiplying the second row by 1/15 replaces the second element by 1. To replace the third element by 0, we add row 2 to row 3. We then obtain

$$\mathbf{A}_3 = \begin{pmatrix} 1 & 0 & 1 & : & 1 & -0.2 & 0 \\ 0 & 1 & 1/3 & : & -1/6 & 1/15 & 0 \\ 0 & 0 & 2 & : & -5.5 & 1 & 1 \end{pmatrix} \begin{matrix} \mathbf{r}_1 - (3/15)\mathbf{r}_2 \\ (1/15)\mathbf{r}_2 \\ \mathbf{r}_3 + \mathbf{r}_2 \end{matrix}$$

Finally, we work on the third column. We multiply row 3 by −0.5 and add the result to row 1; we multiply row 3 by −1/6 and add the result to row 2; and we multiply row 3 by 0.5. We obtain

$$\mathbf{A}_4 = \begin{pmatrix} 1 & 0 & 0 & : & 3.75 & -0.7 & -1/2 \\ 0 & 1 & 0 & : & 0.75 & -0.1 & -1/6 \\ 0 & 0 & 1 & : & -2.75 & 0.5 & 1/2 \end{pmatrix} \begin{matrix} \mathbf{r}_1 - 0.5\mathbf{r}_3 \\ \mathbf{r}_2 - (1/6)\mathbf{r}_3 \\ 0.5\mathbf{r}_3 \end{matrix}.$$

Now, the matrix **A** in (6.34) has been replaced by **I** in $\mathbf{A}_4$. Therefore, the matrix in $\mathbf{A}_4$ which has replaced **I** in (6.34) must be the inverse of **A**, i.e.,

$$\mathbf{A}^{-1} = \begin{pmatrix} 3.75 & -0.7 & -1/2 \\ 0.75 & -0.1 & -1/6 \\ -2.75 & 0.5 & 1/2 \end{pmatrix}.$$

To check, we have

$$\mathbf{AA}^{-1} = \begin{pmatrix} 2 & 6 & 4 \\ 5 & 30 & 15 \\ 6 & 3 & 9 \end{pmatrix} \begin{pmatrix} 3.75 & -0.7 & -1/2 \\ 0.75 & -0.1 & -1/6 \\ -2.75 & 0.5 & 1/2 \end{pmatrix} = \begin{pmatrix} 1 & 0 & 0 \\ 0 & 1 & 0 \\ 0 & 0 & 1 \end{pmatrix} = \mathbf{I}_3.$$

The reader should verify that $\mathbf{A}^{-1}\mathbf{A}$ is also equal to $\mathbf{I}_3$.

When the inverse does not exist. Let us now try to find the inverse of

$$\mathbf{B} = \begin{pmatrix} 2 & 6 & 4 \\ 5 & 30 & 15 \\ 7 & 36 & 19 \end{pmatrix}. \tag{6.36}$$

The augmented matrix in this case is

$$\mathbf{B}_1 = (\mathbf{B} \ : \ \mathbf{I}) = \begin{pmatrix} 2 & 6 & 4 : 1 & 0 & 0 \\ 5 & 30 & 15 : 0 & 1 & 0 \\ 7 & 36 & 19 : 0 & 0 & 1 \end{pmatrix}.$$

We use row operations to reduce the first column of **B** to the first column of **I**. We obtain

$$\mathbf{B}_2 = \begin{pmatrix} 1 & 3 & 2 : & 0.5 & 0 & 0 \\ 0 & 15 & 5 : & -2.5 & 1 & 0 \\ 0 & 5 & 5 : & -3.5 & 0 & 1 \end{pmatrix} \begin{matrix} 0.5\mathbf{r}_1 \\ \mathbf{r}_2 - 2.5\mathbf{r}_1 \\ \mathbf{r}_3 - 3.5\mathbf{r}_1 \end{matrix}.$$

The reduction of the second column gives

$$\mathbf{B}_3 = \begin{pmatrix} 1 & 0 & 0 & : & 1 & -0.2 & 0 \\ 0 & 1 & 1/3 & : & -1/6 & 1/15 & 0 \\ 0 & 0 & 0 & : & -1 & -1 & 1 \end{pmatrix} \begin{matrix} \mathbf{r}_1 - (1/3)\mathbf{r}_2 \\ (1/15)\mathbf{r}2 \\ \mathbf{r}_3 - \mathbf{r}_2 \end{matrix}$$

So far so good! But an examination of $\mathbf{B}_3$ shows that we won't be able to reduce the third column. Therefore, the above method for finding the inverse of **B** fails. The reason for this, however, is the fact that the inverse of **B** does not exist. To see this, let us compute the rank of **B**. We need to compute a row-echelon form of **B**, but we have already done that because the first three columns of $\mathbf{B}_3$ are in fact a row-echelon form of **B**. There are two nonzero rows in the echelon form of **B**, which indicates that $r(\mathbf{B}) = 2$. The matrix **B** contains linearly dependent vectors and is, therefore, rank-deficient (an examination of **B** shows that the third row is equal to the sum of the first two rows). Thus, by Result 6.1, **B** is singular and its inverse does not exist.

Exercises

6.1. Gasoline consumption depends on whether a car is driven on city streets or on highways. Each of three cars has been driven for 20 miles on city streets and 20 miles on highways. The gasoline consumption (number of gallons each car has consumed) is given in the following matrix:

$$\mathbf{C} = \begin{pmatrix} 1.0 & 0.8 \\ 1.6 & 1.0 \\ 2.0 & 1.6 \end{pmatrix}.$$

(a) Which column of **C** would represent city streets and which would represent highways?
(b) Compute the matrix **G** (symbolically in terms of **C** and numerically) whose elements are the gasoline consumption in terms of gallons per mile.
(c) Compute the matrix **M** that contains the element-wise division of **G**.
(d) Interpret the elements of **M**.
(e) Let $\mathbf{a} = (1/2)(1 \quad 1)^{\mathrm{T}}$ and $\mathbf{b} = (1/3)(1 \quad 1 \quad 1)^{\mathrm{T}}$. Interpret the following matrix products: $\mathbf{Ga}$, $\mathbf{Ma}$, $\mathbf{G}^{\mathrm{T}}\mathbf{b}$, and $\mathbf{M}^{\mathrm{T}}\mathbf{b}$.

6.2. Verify Result 6.2 (c)–(f) using $\mathbf{A} = \begin{pmatrix} 1 & 2 \\ 5 & 4 \end{pmatrix}$ and $\mathbf{B} = \begin{pmatrix} 1 & 2 \\ 2 & 8 \end{pmatrix}$.

6.3. Compute the inverse of each of the following matrices, where possible, and check your calculations by multiplying the matrix by its inverse and obtaining **I**. If a matrix does not have an inverse, find one of its g-inverses.

$$\mathbf{A} = \begin{pmatrix} 1 & 2 \\ 2 & 6 \end{pmatrix}, \quad \mathbf{B} = \begin{pmatrix} 1 & 3 \\ 2 & -6 \end{pmatrix}, \quad \mathbf{C} = \begin{pmatrix} 2 & 1 & 2 \\ 1 & 0 & 1 \end{pmatrix}, \quad \mathbf{D} = \begin{pmatrix} 4 & 0 & 0 \\ 0 & 1 & 0 \\ 0 & 0 & -1 \end{pmatrix}.$$

6.4. Verify that one of the following two matrices is the inverse of the other, then compute $(\mathbf{A}^{-1})^{-1}$, $(\mathbf{A}^{\mathrm{T}})^{-1}$, and $(\mathbf{A}^{-1})^{\mathrm{T}}$:

$$\mathbf{A} = \begin{pmatrix} 1 & 1 & -1 \\ -1 & 1 & 1 \\ 1 & -1 & 1 \end{pmatrix} \quad \text{and} \quad \mathbf{B} = \begin{pmatrix} 0.5 & 0.0 & 0.5 \\ 0.5 & 0.5 & 0.0 \\ 0.0 & 0.5 & 0.5 \end{pmatrix}.$$

6.5. Let $\mathbf{P} = \begin{pmatrix} \mathbf{A} & \mathbf{y} \\ \mathbf{x}^T & 1 \end{pmatrix}$, where $\mathbf{A}$ is as given in Exercise 6.3 and $\mathbf{y} = 2\mathbf{x} = \begin{pmatrix} 0 \\ 2 \end{pmatrix}$.

(a) Compute $\mathbf{P}^{-1}$ using $\mathbf{A}^{-1}$ found in Exercise 6.3, then verify that $\mathbf{P}^{-1}\mathbf{P} = \mathrm{I}$.
(b) Compute det($\mathbf{P}$) using Result 6.7.
(c) Compute det($\mathbf{P}$) using Result 6.7, but after replacing $\mathbf{A}$ by $\mathbf{B}$ in Exercise 6.3.

6.6. Partition $\mathbf{P} = \begin{pmatrix} 1 & 2 & 1 \\ 2 & 4 & 0 \\ 1 & 0 & 1 \end{pmatrix}$ as in (6.5), in any way you like; then:

(a) Compute det($\mathbf{P}$) using Result 6.7.
(b) Find $\mathbf{P}^{-1}$ using Result 6.4, then verify that $\mathbf{P}^{-1}\mathbf{P} = \mathrm{I}$.

6.7. Given $\mathbf{P} = \begin{pmatrix} 1 & 2 & 1 \\ 4 & 5 & 1 \\ 0 & 1 & 1 \end{pmatrix}$,

(a) Show that r($\mathbf{P}$) = 2, hence $\mathbf{P}$ is singular.
(b) Partition $\mathbf{P}$ as in (6.5), but in any way you like; then, using Result 6.7, show that det($\mathbf{P}$) = 0, hence $\mathbf{P}$ is singular.
(c) Using the partitioned $\mathbf{P}$ and Result 6.4, try to compute $\mathbf{P}^{-1}$. Your calculations should break down at some point. Indicate where and why.

6.8. The following are square nonsingular matrices:

$$\mathbf{P} = \begin{pmatrix} \mathbf{A} & \mathbf{0} \\ \mathbf{0} & \mathbf{B} \end{pmatrix} \quad \text{and} \quad \mathbf{Q} = \begin{pmatrix} \mathbf{P} & \mathbf{0} \\ \mathbf{0} & \mathbf{C} \end{pmatrix} = \begin{pmatrix} \mathbf{A} & \mathbf{0} & \mathbf{0} \\ \mathbf{0} & \mathbf{B} & \mathbf{0} \\ \mathbf{0} & \mathbf{0} & \mathbf{C} \end{pmatrix},$$

where each of the null matrices is of an appropriate order. Matrices like $\mathbf{P}$ and $\mathbf{Q}$ are called *block-diagonal* matrices. Using Result 6.4, show that

$$\mathbf{P}^{-1} = \begin{pmatrix} \mathbf{A}^{-1} & \mathbf{0} \\ \mathbf{0} & \mathbf{B}^{-1} \end{pmatrix} \quad \text{and} \quad \mathbf{Q}^{-1} = \begin{pmatrix} \mathbf{P}^{-1} & \mathbf{0} \\ \mathbf{0} & \mathbf{C}^{-1} \end{pmatrix} = \begin{pmatrix} \mathbf{A}^{-1} & \mathbf{0} & \mathbf{0} \\ \mathbf{0} & \mathbf{B}^{-1} & \mathbf{0} \\ \mathbf{0} & \mathbf{0} & \mathbf{C}^{-1} \end{pmatrix}.$$

Hence, by repeated applications of Result 6.4, one can show that the inverse of a block-diagonal matrix is block-diagonal with the inverses of the submatrices on the corresponding diagonal blocks.

6.9. Let $\mathbf{Q} = \begin{pmatrix} 1 & 2 \\ 5 & 4 \end{pmatrix}$, $\mathbf{y} = 2\mathbf{x} = \begin{pmatrix} 0 \\ 2 \end{pmatrix}$, and $s = 2$.

(a) Compute $\mathbf{R} = (\mathbf{Q} + 2\mathbf{y}\mathbf{x}^{\mathrm{T}})^{-1}$.
(b) Compute $\mathbf{R}^{-1}$ using (6.4).
(c) Compute $\mathbf{R}^{-1}$ using Result 6.5.

6.10. Find any matrix $\mathbf{A}$ such that $\mathbf{A} \neq \mathbf{I}$ and $\mathbf{A} = \mathbf{A}^{-1}$.

6.11. (a) Show that if $\det(\mathbf{A}) \neq 0$ and $\mathbf{AB} = \mathbf{AC}$, then $\mathbf{B} = \mathbf{C}$.
(b) Find three matrices $\mathbf{A}$, $\mathbf{B}$, and $\mathbf{C}$ such that $\mathbf{AB} = \mathbf{AC}$ but $\mathbf{B} \neq \mathbf{C}$.
(c) Either prove that any matrix with zero diagonal elements is singular, or find a counterexample.

6.12. Let $\mathbf{A}$ and $\mathbf{B}$ be orthogonal matrices of the same order.
(a) Give an example of $\mathbf{A}$ and $\mathbf{B}$.
(b) Compute the inverses of your matrices in (a).
(c) Show that the determinant of any orthogonal matrix is either 1 or –1.
(d) Show that $\mathbf{AB}$ is an orthogonal matrix.
(e) Show that if $\mathbf{A}$ and $\mathbf{B}$ are also symmetric, then $\mathbf{AB} = \mathbf{BA}$.

6.13. Use the algorithm in the appendix to this chapter to compute the inverse of each of the following matrices, where possible:

$$\mathbf{A} = \begin{pmatrix} 2 & 0 & 0 \\ 6 & 2 & 0 \\ 0 & 1 & 1 \end{pmatrix}, \quad \mathbf{B} = \begin{pmatrix} 2 & 3 & 0 \\ 0 & 1 & 5 \\ 0 & 0 & 1 \end{pmatrix}, \quad \mathbf{C} = \begin{pmatrix} 2 & 3 & 0 \\ 0 & 0 & 5 \\ 0 & 0 & 1 \end{pmatrix}, \quad \mathbf{D} = \begin{pmatrix} 2 & 0 & 4 \\ 0 & 1 & 0 \\ 1 & 1 & 0 \end{pmatrix}.$$

6.14. Let $\mathbf{P} = \begin{pmatrix} \mathbf{I} & \mathbf{A} \\ \mathbf{0} & \mathbf{B} \end{pmatrix}$ and $\mathbf{Q} = \begin{pmatrix} \mathbf{I} & \mathbf{0} \\ \mathbf{C} & \mathbf{B} \end{pmatrix}$, where $\mathbf{B}$ is a square matrix. Show that $\det(\mathbf{P}) = \det(\mathbf{Q}) = \det(\mathbf{B})$.

6.15. What are the orders of the matrices $\mathbf{G}_{11}$, $\mathbf{G}_{12}$, $\mathbf{G}_{21}$, and $\mathbf{G}_{22}$ in (6.16)?

6.16. In each of the following cases, $\mathbf{G}$ is a g-inverse of $\mathbf{A}$. Find how many of the conditions in (6.24)–(6.27) $\mathbf{G}$ satisfies.

(a) $\mathbf{A} = \begin{pmatrix} 1 & 1 & 1 \\ 0 & 0 & 0 \\ 1 & 1 & 1 \\ 1 & 0 & 1 \end{pmatrix}$, $\quad \mathbf{G} = \begin{pmatrix} -1.0 & -1 & -1.0 & 0 \\ 0.5 & 0 & 0.5 & -1 \\ 1.0 & 1 & 1.0 & 1 \end{pmatrix}$.

(b) $\mathbf{A} = \begin{pmatrix} 1 & 1 & 1 \\ 1 & 0 & 1 \\ 0 & 0 & 0 \\ 1 & 1 & 1 \end{pmatrix}$, $\mathbf{G} = \begin{pmatrix} -1.0 & 0 & -1 & -1.0 \\ 0.5 & -1 & 0 & 0.5 \\ 1.0 & 1 & 1 & 1.0 \end{pmatrix}$.

(c) $\mathbf{A} = \begin{pmatrix} 1 & 1 & 1 \\ 1 & 0 & 1 \\ 0 & 0 & 0 \\ 1 & 1 & 1 \end{pmatrix}$, $\mathbf{G} = \begin{pmatrix} 0 & 1 & 0 & 0 \\ 1 & -1 & 0 & 0 \\ 0 & 0 & 0 & 0 \end{pmatrix}$.

(d) $\mathbf{A} = \begin{pmatrix} 1 & 1 & 1 \\ 1 & 0 & 0 \\ 1 & 1 & 1 \\ 1 & 0 & 1 \end{pmatrix}$, $\mathbf{G} = \begin{pmatrix} 0.0 & 1 & 0.0 & 0 \\ 0.5 & 0 & 0.5 & -1 \\ 0.0 & -1 & 0.0 & 1 \end{pmatrix}$.

(e) $\mathbf{A} = \begin{pmatrix} 1 & 1 & 1 \\ 1 & 1 & 1 \\ 1 & 0 & 0 \\ 1 & 0 & 1 \end{pmatrix}$, $\mathbf{G} = \begin{pmatrix} 0.0 & 0.0 & 1 & 0 \\ 0.5 & 0.5 & 0 & -1 \\ 0.0 & 0.0 & -1 & 1 \end{pmatrix}$.

(f) $\mathbf{A} = \begin{pmatrix} 1 & 2 & 1 \\ 0 & 0 & 0 \\ 1 & 2 & 1 \\ 0 & 0 & 0 \end{pmatrix}$, $\mathbf{G} = \begin{pmatrix} -0.5 & -1.0 & -0.5 & -1 \\ 0.0 & 0.0 & 0.0 & 0 \\ 1.0 & 1.0 & 1.0 & 1 \end{pmatrix}$.

6.17. Compute a g-inverse of each of the following matrices:

$$\mathbf{A} = \begin{pmatrix} 1 & 1 & 0 \\ 1 & 0 & 1 \end{pmatrix}, \quad \mathbf{B} = \begin{pmatrix} 0 & 1 & 0 \\ 1 & 0 & 1 \end{pmatrix}, \quad \mathbf{C} = \begin{pmatrix} 2 \\ 4 \end{pmatrix},$$

$$\mathbf{D} = \begin{pmatrix} 1 & 1 \\ 1 & 1 \\ 1 & 1 \end{pmatrix}, \quad \mathbf{E} = \begin{pmatrix} 0 & 1 \\ 0 & 1 \\ 0 & 1 \end{pmatrix}, \quad \mathbf{F} = (1 \quad 2).$$

6.18. Compute another g-inverse of each of the matrices in Exercise 6.17.

6.19. Compute a g-inverse of the transpose of each of the matrices in Exercise 6.17.

6.20. Find the Moore-Penrose inverse for each of the following matrices:

$$\mathbf{A} = \begin{pmatrix} 1 & 1 \\ 1 & 2 \\ 1 & 3 \end{pmatrix} \quad \text{and} \quad \mathbf{B} = \begin{pmatrix} 0 \\ 1 \\ 1 \end{pmatrix}.$$

6.21. Verify Result 6.10 using the matrix **A** in Exercise 6.16(a).

6.22. Verify Result 6.11 using the matrix **A** in Exercise 6.16(d).

6.23. Let $\mathbf{A}^{\mathrm{T}} = \begin{pmatrix} 1 & 1 & 1 \\ 0 & 1 & 0 \end{pmatrix}$, then:

(a) Use the full-rank factorization result in (6.31) to find **B** and **C** such that $\mathbf{A} = \mathbf{BC}$.
(b) Use Result 6.12 to compute a g-inverse of **A**.

6.24. Let $\mathbf{A}^{\mathrm{T}} = \begin{pmatrix} 0 & 0 & 1 \\ 0 & 1 & 0 \end{pmatrix}$, then:

(a) Use the full-rank factorization result in (6.31) to find **B** and **C** such that $\mathbf{A} = \mathbf{BC}$.
(b) Use Result 6.12 to compute a g-inverse of **A**.

6.25. Let **A** and **B** be square matrices.
(a) Show that if **A** is orthogonal, then $\mathrm{tr}(\mathbf{A}^{\mathrm{T}}\mathbf{BA}) = \mathrm{tr}(\mathbf{B})$.
(b) Show that if **A** is orthogonal, then $\det(\mathbf{A}^{\mathrm{T}}\mathbf{BA}) = \det(\mathbf{B})$.
(c) Show that if **A** is orthogonal and **B** is idempotent, then $\mathbf{A}^{\mathrm{T}}\mathbf{BA}$ is idempotent.

7 Linear Transformation

7.1. Introduction

Data do not always come in a form suitable for our needs and must be converted to the desired form. For example, suppose that temperature is measured using the Celsius scale and we wish to express it in the Fahrenheit scale. Let F denote the temperature in terms of the Fahrenheit scale and let C denote the same temperature using the Celsius scale. The formula that relates C to F is

$$F = 1.8\ C + 32. \tag{7.1}$$

Thus, for example, if the temperature is 20° Celsius, the equivalent temperature in Fahrenheit is 68°. Formula (7.1) enables us to *transform* or *reexpress* the temperature from one scale to another.

We may think of a transformation as a function that takes inputs and produces outputs. In the example above, the temperature in degrees Celsius is the input and the output is the corresponding temperature in degrees Fahrenheit. Transformations can be categorized into two types: linear transformations and nonlinear transformations. The formula in (7.1) is an example of a linear transformation, while the quadratic function

$$y = ax^2 + bx + c \tag{7.2}$$

is an example of a nonlinear transformation (a quadratic transformation in this case). Here x is the input and y is the output. We shall see other examples of nonlinear transformations in the remaining chapters.

In this chapter we concentrate on linear transformations. In Section 7.2, we define linear transformations and give some numerical examples. In Sections 7.3–7.5, we discuss three types of linear transformations: orthogonal transformations, oblique transformations, and orthogonal projections. Section 7.6 classifies any linear transformation as either a singular or a nonsingular transformation.

7.2. Definition

Let $\mathbf{x}$ and $\mathbf{c}$ be vectors of dimension $n \times 1$ and $\mathbf{A}$ be a matrix of dimension $m \times n$. Any transformation of a vector $\mathbf{x}$ to a vector $\mathbf{y}$ that can be written as

$$\mathbf{y} = \mathbf{A}(\mathbf{x} - \mathbf{c}) \tag{7.3}$$

is called a *linear transformation*. A linear transformation is said to be *nonsingular* if

A is nonsingular, and *singular* if **A** is singular. Note that the order of **y**, which is $m \times 1$, may be different from that of **x**. Here **A** is referred to as the *transformation matrix* and, for reasons to be explained shortly, **c** is called the *shift* vector.

For example, the reader can verify that (7.1) may be written as

$$F = \frac{9}{5}C + 32 = \frac{9}{5}\left(C - \frac{-160}{9}\right).$$

By comparison with (7.3), we see that $\mathbf{A} = 9/5$ and $\mathbf{c} = -160/9$. Thus, (7.1) is a linear transformation. Note that in this example, $m = n = 1$, and hence **A** is a 1×1 matrix and **c** is a 1×1 vector.

Let us now consider some examples in two dimensions that will also help us interpret the effects of a linear transformation. Consider the vectors

$$\mathbf{v}_1 = \begin{pmatrix} 4 \\ 2 \end{pmatrix}, \quad \mathbf{v}_2 = \begin{pmatrix} 4 \\ 4 \end{pmatrix}, \quad \text{and} \quad \mathbf{v}_3 = \begin{pmatrix} -1 \\ 2 \end{pmatrix}, \tag{7.4}$$

and suppose we wish to transform each of these vectors using

$$\mathbf{A} = \frac{1}{\sqrt{5}}\begin{pmatrix} 2 & 1 \\ -1 & 2 \end{pmatrix} \quad \text{and} \quad \mathbf{c} = \begin{pmatrix} 0 \\ 0 \end{pmatrix}. \tag{7.5}$$

From (7.3), we obtain

$$\mathbf{w}_1 = \mathbf{A}(\mathbf{v}_1 - \mathbf{c}) = \frac{1}{\sqrt{5}}\begin{pmatrix} 2 & 1 \\ -1 & 2 \end{pmatrix}\left\{\begin{pmatrix} 4 \\ 2 \end{pmatrix} - \begin{pmatrix} 0 \\ 0 \end{pmatrix}\right\} = \frac{1}{\sqrt{5}}\begin{pmatrix} 10 \\ 0 \end{pmatrix} = \begin{pmatrix} 2\sqrt{5} \\ 0 \end{pmatrix}.$$

Therefore, $\mathbf{w}_1$ is a linear transformation of $\mathbf{v}_1$. Similarly,

$$\mathbf{w}_2 = \mathbf{A}(\mathbf{v}_2 - \mathbf{c}) = \frac{1}{\sqrt{5}}\begin{pmatrix} 2 & 1 \\ -1 & 2 \end{pmatrix}\left\{\begin{pmatrix} 4 \\ 4 \end{pmatrix} - \begin{pmatrix} 0 \\ 0 \end{pmatrix}\right\} = \frac{1}{\sqrt{5}}\begin{pmatrix} 12 \\ 4 \end{pmatrix};$$

and

$$\mathbf{w}_3 = \mathbf{A}(\mathbf{v}_3 - \mathbf{c}) = \frac{1}{\sqrt{5}}\begin{pmatrix} 2 & 1 \\ -1 & 2 \end{pmatrix}\left\{\begin{pmatrix} -1 \\ 2 \end{pmatrix} - \begin{pmatrix} 0 \\ 0 \end{pmatrix}\right\} = \frac{1}{\sqrt{5}}\begin{pmatrix} 0 \\ 5 \end{pmatrix} = \begin{pmatrix} 0 \\ \sqrt{5} \end{pmatrix}.$$

Thus, using **A** and **c** in (7.5) the vectors $\mathbf{v}_1$, $\mathbf{v}_2$, and $\mathbf{v}_3$ in (7.4) are transformed to

$$\mathbf{w}_1 = \begin{pmatrix} 2\sqrt{5} \\ 0 \end{pmatrix}, \quad \mathbf{w}_2 = \frac{1}{\sqrt{5}}\begin{pmatrix} 12 \\ 4 \end{pmatrix}, \quad \text{and} \quad \mathbf{w}_3 = \begin{pmatrix} 0 \\ \sqrt{5} \end{pmatrix}, \tag{7.6}$$

respectively.

To see the effects of this linear transformation, let us graph $\mathbf{v}_1$, $\mathbf{v}_2$, and $\mathbf{v}_3$ in a two-dimensional space (see Chapter 4). The three points representing these vectors are shown in Figure 7.1. The coordinates of each of these points are measured with reference to the horizontal and vertical axes, hence labeled x_1 and x_2.

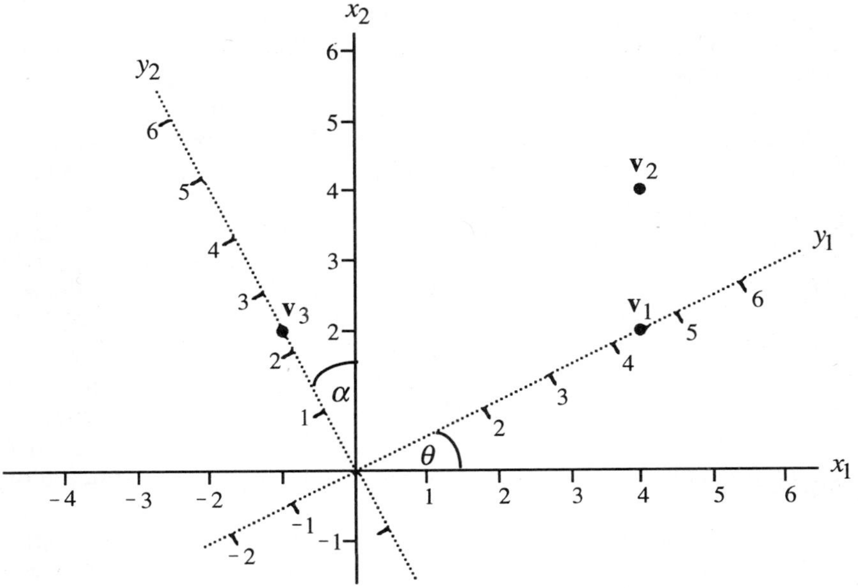

Figure 7.1. An example of a linear transformation of the vectors $\mathbf{v}_1$, $\mathbf{v}_2$, and $\mathbf{v}_3$ in (7.4). This linear transformation, which is given by $\mathbf{A}$ and $\mathbf{c}$ in (7.5), results in a rotation of the original set of axes (x_1, x_2) to a new set of axes (y_1, y_2). The vectors $\mathbf{w}_1$, $\mathbf{w}_2$, and $\mathbf{w}_3$ in (7.6) contain the coordinates of the three graphed points, measured with reference to the (y_1, y_2) axes.

Now suppose that, instead of using the traditional x_1 and x_2 axes, we wish to measure the coordinates of the graphed points with reference to the y_1 and y_2 axes in Figure 7.1. (The new axes are marked by the dotted lines.) Note that the vectors $\mathbf{v}_1$ and $\mathbf{v}_3$ coincide with the y_1- and y_2-axis, respectively. Let us measure the angle θ between the y_1 and x_1 axes. The y_1-axis can be represented by $\mathbf{v}_1$ (or any nonzero multiple of $\mathbf{v}_1$) and the x_1-axis can be represented by the first unit vector, $\mathbf{u}_1 = (1 \quad 0)^{\mathrm{T}}$ (or any nonzero multiple of $\mathbf{u}_1$). Thus, the angle between the y_1 and y_2 axes is the same as the angle between $\mathbf{v}_1$ and $\mathbf{u}_1$. By Result 4.1, we have

$$\cos\theta = \frac{\mathbf{u}_1^{\mathrm{T}}\mathbf{v}_1}{\|\mathbf{u}_1\|\times\|\mathbf{v}_1\|} = \frac{4}{(1)\sqrt{20}} = \frac{2}{\sqrt{5}},$$

which implies that θ is approximately 27°. Thus, the y_1-axis makes a 27° angle with the x_1-axis and a 63° angle with the x_2-axis. Similarly, the angle α between the y_2 and x_2 axes is the same as the angle between $\mathbf{v}_3$ and $\mathbf{u}_2 = (0 \quad 1)^{\mathrm{T}}$ and can be found from

$$\cos\alpha = \frac{\mathbf{u}_2^{\mathrm{T}}\mathbf{v}_3}{\|\mathbf{u}_2\|\times\|\mathbf{v}_3\|} = \frac{2}{(1)\sqrt{5}} = \frac{2}{\sqrt{5}},$$

which implies that α is approximately 27° and that the y_2-axis makes a 27° angle with the x_2-axis. Therefore, the new set of axes is obtained by rotating the original set of axes an angle of 27° counterclockwise.

A question of interest is, What are the coordinates of each of the points in Figure 7.1 measured with respect to the new set of axes? The new coordinates are found by measuring the distance from the origin to each point, with reference to the new set of axes. Thus, it can be seen from Figure 7.1 that the new coordinates of the point $\mathbf{v}_1$ are $(2\sqrt{5}, 0)$, that is, $y_1 = 2\sqrt{5}$ and $y_2 = 0$. These coordinates are nothing but the elements of the vector $\mathbf{w}_1$ in (7.6). For the point $\mathbf{v}_2$, the new coordinates are $(12/\sqrt{5}, 4/\sqrt{5})$, that is, $y_1 = 12/\sqrt{5}$ and $y_2 = 4/\sqrt{5}$, which are the same as the elements of $\mathbf{w}_2$. Similarly, the new coordinates of $\mathbf{v}_3$ are $(0, \sqrt{5})$, which are the same as $\mathbf{w}_3$.

In this example, the linear transformation can be interpreted simply as an *axes rotation*. In other words, the location of the points in the space remains unchanged but we simply change the angle from which we look at the points; the origin remains the same but the axes are rotated counterclockwise by a 27° angle. In Section 7.3, we shall see why the matrix $\mathbf{A}$ in (7.5) gave rise to a counterclockwise rotation of a 27° angle.

Let us examine another example of a linear transformation which is slightly different from the one above. Suppose we wish to transform each of the vectors in (7.4), but now using

$$\mathbf{A} = \frac{1}{\sqrt{5}}\begin{pmatrix} 2 & 1 \\ -1 & 2 \end{pmatrix} \quad \text{and} \quad \mathbf{c} = \begin{pmatrix} 2 \\ 2 \end{pmatrix}. \tag{7.7}$$

That is, the transformation matrix $\mathbf{A}$ remains the same but we change the shift vector $\mathbf{c}$. Using (7.3), we obtain

$$\mathbf{r}_1 = \mathbf{A}(\mathbf{v}_1 - \mathbf{c}) = \frac{1}{\sqrt{5}}\begin{pmatrix} 2 & 1 \\ -1 & 2 \end{pmatrix}\left\{\begin{pmatrix} 4 \\ 2 \end{pmatrix} - \begin{pmatrix} 2 \\ 2 \end{pmatrix}\right\} = \frac{1}{\sqrt{5}}\begin{pmatrix} 4 \\ -2 \end{pmatrix}.$$

Similarly,

$$\mathbf{r}_2 = \mathbf{A}(\mathbf{v}_2 - \mathbf{c}) = \frac{1}{\sqrt{5}}\begin{pmatrix} 2 & 1 \\ -1 & 2 \end{pmatrix}\left\{\begin{pmatrix} 4 \\ 4 \end{pmatrix} - \begin{pmatrix} 2 \\ 2 \end{pmatrix}\right\} = \frac{1}{\sqrt{5}}\begin{pmatrix} 6 \\ 2 \end{pmatrix},$$

and

$$\mathbf{r}_3 = \mathbf{A}(\mathbf{v}_3 - \mathbf{c}) = \frac{1}{\sqrt{5}}\begin{pmatrix} 2 & 1 \\ -1 & 2 \end{pmatrix}\left\{\begin{pmatrix} -1 \\ 2 \end{pmatrix} - \begin{pmatrix} 2 \\ 2 \end{pmatrix}\right\} = \frac{1}{\sqrt{5}}\begin{pmatrix} -6 \\ 3 \end{pmatrix}.$$

Thus, using $\mathbf{A}$ and $\mathbf{c}$ in (7.7), the vectors $\mathbf{v}_1$, $\mathbf{v}_2$, and $\mathbf{v}_3$ in (7.4) are transformed to

$$\mathbf{r}_1 = \frac{1}{\sqrt{5}}\begin{pmatrix} 4 \\ -2 \end{pmatrix}, \quad \mathbf{r}_2 = \frac{1}{\sqrt{5}}\begin{pmatrix} 6 \\ 2 \end{pmatrix}, \quad \text{and} \quad \mathbf{r}_3 = \frac{1}{\sqrt{5}}\begin{pmatrix} -6 \\ 3 \end{pmatrix}, \tag{7.8}$$

respectively. Whereas in the previous example the new vectors $\mathbf{w}_1$, $\mathbf{w}_2$, and $\mathbf{w}_3$ represent the coordinates of the points measured with reference to the new set of axes, in this example, $\mathbf{r}_1$, $\mathbf{r}_2$, and $\mathbf{r}_3$ do not. The reason for this is that in the previous example we have kept the origin unchanged and we have only rotated the axes, but in the current example we have rotated the axes and shifted the origin from the point $\mathbf{c} = (0, 0)^T$ to the point $\mathbf{c} = (2, 2)^T$. This is the reason that $\mathbf{c}$ in (7.3) is called the shift vector.

Figure 7.2 is the graph of $\mathbf{v}_1$, $\mathbf{v}_2$, and $\mathbf{v}_3$ as in Figure 7.1, but here we first shifted the origin from the point (0, 0) to the point (2, 2), then rotated the axes counterclockwise by a 27° angle. Now, when we measure the locations of the points $\mathbf{v}_1$, $\mathbf{v}_2$, and $\mathbf{v}_3$ with respect to this new set of axes, we see that their coordinates are indeed the elements of the vectors $\mathbf{r}_1$, $\mathbf{r}_2$, and $\mathbf{r}_3$, respectively.

In the first of the two preceding examples, linear transformation results only in a rotation. The second example involves both a shift of the origin and a rotation of the axes. Let us now consider a third example in which the transformation involves a shift of the origin but no axes rotation. In this example we transform the vectors in (7.4) using

$$\mathbf{A} = \begin{pmatrix} 1 & 0 \\ 0 & 1 \end{pmatrix} \quad \text{and} \quad \mathbf{c} = \begin{pmatrix} 2 \\ 2 \end{pmatrix}. \tag{7.9}$$

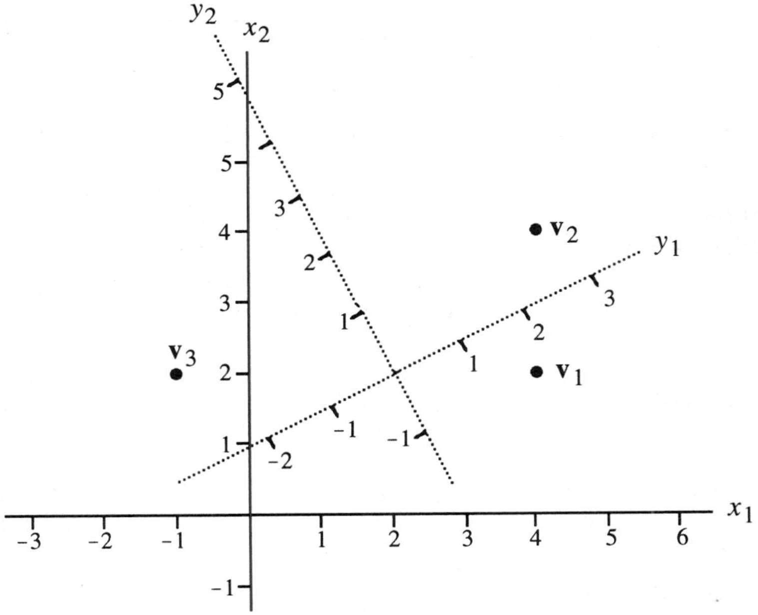

Figure 7.2. An example of a linear transformation of the vectors $\mathbf{v}_1$, $\mathbf{v}_2$, and $\mathbf{v}_3$ in (7.4). This linear transformation, which is given by $\mathbf{A}$ and $\mathbf{c}$ in (7.7), results in a shift of the origin and a rotation of the axes (x_1, x_2) to a new set of axes (y_1, y_2). The vectors $\mathbf{r}_1$, $\mathbf{r}_2$, and $\mathbf{r}_3$ in (7.8) contain the coordinates of the points measured with reference to the (y_1, y_2) axes.

The reader can verify that

$$\mathbf{z}_1 = \mathbf{A}(\mathbf{v}_1 - \mathbf{c}) = \left\{\begin{pmatrix}4\\2\end{pmatrix} - \begin{pmatrix}2\\2\end{pmatrix}\right\} = \begin{pmatrix}2\\0\end{pmatrix}.$$

Similarly,

$$\mathbf{z}_2 = \mathbf{A}(\mathbf{v}_2 - \mathbf{c}) = \left\{\begin{pmatrix}4\\4\end{pmatrix} - \begin{pmatrix}2\\2\end{pmatrix}\right\} = \begin{pmatrix}2\\2\end{pmatrix},$$

and

$$\mathbf{z}_3 = \mathbf{A}(\mathbf{v}_3 - \mathbf{c}) = \left\{\begin{pmatrix}-1\\2\end{pmatrix} - \begin{pmatrix}2\\2\end{pmatrix}\right\} = \begin{pmatrix}-3\\0\end{pmatrix}.$$

Thus, using **A** and **c** in (7.9) the vectors $\mathbf{v}_1$, $\mathbf{v}_2$, and $\mathbf{v}_3$ are transformed to

$$\mathbf{z}_1 = \begin{pmatrix}2\\0\end{pmatrix}, \quad \mathbf{z}_2 = \begin{pmatrix}2\\2\end{pmatrix}, \quad \text{and} \quad \mathbf{z}_3 = \begin{pmatrix}-3\\0\end{pmatrix}, \tag{7.10}$$

respectively. The graph of this transformation is shown in Figure 7.3. The origin is shifted from (0, 0) to (2, 2) but the axes remain unrotated; the y_1-axis is still horizontal and the y_2-axis is still vertical. As we can see from Figure 7.3, the vectors $\mathbf{z}_1$, $\mathbf{z}_2$, and $\mathbf{z}_3$ contain the coordinates of the three points measured with respect to this new set of axes.

*7.3. Orthogonal Rotation

The linear transformations in Figures 7.1–7.3 have one thing in common: In each transformation, the new axes are perpendicular to one another. This type of linear transformation is called an *orthogonal transformation* or *orthogonal rotation* because the new axes are obtained by rigidly and simultaneously rotating the original axes, keeping the right angle between them unchanged. Like the original axes, the new axes remain mutually perpendicular or orthogonal.

How do we know whether a given linear transformation is orthogonal? How do we create an orthogonal linear transformation if one is desired? The answers to these questions are given below.

Result 7.1. A linear transformation, as in (7.3), is orthogonal if and only if the transformation matrix is orthogonal.[1]

By Result 7.1, we can then determine whether a linear transformation is orthogonal simply by checking whether the transformation matrix is orthogonal. Thus, for example, the reader can verify that the transformation matrices in (7.5), (7.7), and (7.9) are orthogonal, and hence each of the transformations in Figures 7.1–7.3 must be orthogonal.

---

1 Recall, from the definition in Chapter 2, that a square matrix **A** is said to be orthogonal if it satisfies $\mathbf{A}^T\mathbf{A} = \mathbf{I}$.

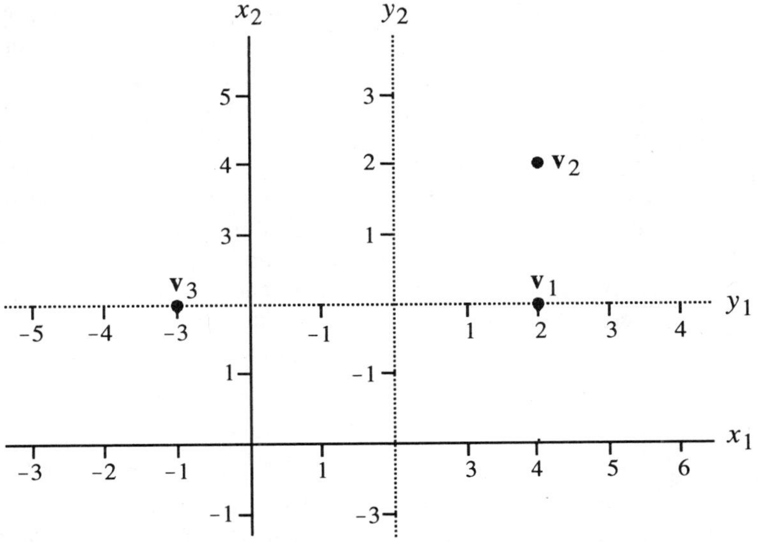

Figure 7.3. An example of a linear transformation of the vectors $\mathbf{v}_1$, $\mathbf{v}_2$, and $\mathbf{v}_3$ in (7.4). This linear transformation, which is given by $\mathbf{A}$ and $\mathbf{c}$ in (7.9), results in a shift of the origin but no rotation of the axes. The vectors $\mathbf{z}_1$, $\mathbf{z}_2$, and $\mathbf{z}_3$ in (7.10) contain the coordinates of the points measured with reference to the (y_1, y_2) axes.

We can also use Result 7.1 to create an orthogonal linear transformation. We simply choose a transformation matrix that is orthogonal. Recall that all vectors in an orthogonal matrix must be orthonormal. We also know from Result 4.6 that the jth element of any orthonormal vector is the direction cosine of the angle between the vector and the jth coordinate axis. Therefore, a transformation matrix that rotates the two axes counterclockwise by an angle θ will have the form

$$\mathbf{A} = \begin{pmatrix} \cos\theta & \sin\theta \\ -\sin\theta & \cos\theta \end{pmatrix}. \tag{7.11}$$

The matrix $\mathbf{A}$ in (7.11), which depends only on θ, is orthogonal because

$$\begin{aligned} \mathbf{A}^{\mathrm{T}}\mathbf{A} &= \begin{pmatrix} \cos\theta & -\sin\theta \\ \sin\theta & \cos\theta \end{pmatrix} \begin{pmatrix} \cos\theta & \sin\theta \\ -\sin\theta & \cos\theta \end{pmatrix} \\ &= \begin{pmatrix} (\cos\theta)^2 + (\sin\theta)^2 & 0 \\ 0 & (\cos\theta)^2 + (\sin\theta)^2 \end{pmatrix} = \begin{pmatrix} 1 & 0 \\ 0 & 1 \end{pmatrix}, \end{aligned}$$

because of the trigonometric identity $(\cos\theta)^2 + (\sin\theta)^2 = 1$.

Thus, by comparing the matrix **A** in (7.5) with that in (7.11), we see, for example, that $a_{11} = 2/\sqrt{5} = \cos\theta$, which implies that θ is approximately 27°. Of course, if we chose some other element of **A**, we would end up with the same result. For example, $a_{12} = 1/\sqrt{5} = \sin\theta$, which again implies that θ is approximately 27°.

Orthogonal linear transformations in higher dimensions work in the same way as in the two-dimensional case. The transformation matrix **A** has to be orthogonal and the element a_{ij} is the cosine of the angle between the ith old coordinate axis and the jth new axis. For example, if x_1, x_2, and x_3 are the original axes in a space of three dimensions and y_1, y_2, and y_3 are the new set of axes, then

a_{11} is the cosine of the angle between x_1 and y_1,
a_{21} is the cosine of the angle between x_2 and y_1,
a_{31} is the cosine of the angle between x_3 and y_1,
a_{12} is the cosine of the angle between x_1 and y_2, and so on.

*7.4. Oblique Rotation

We have seen in the previous section that orthogonal transformations can be interpreted graphically as a simultaneous rigid rotation of the original axes. It might be desirable in some cases to rotate each axis independently from the other axes. For example, we may rotate the x_1-axis by a 30° angle, while rotating the x_2-axis by a 45° angle. Unlike the original axes, the new axes are no longer mutually perpendicular or orthogonal. This type of linear transformation is called an *oblique transformation* or *oblique rotation*. Generally speaking, if the transformation matrix is orthogonal, the transformation or rotation is orthogonal; otherwise it is oblique.

Oblique transformations have many practical applications. Examples of some applications are given in Chapter 11.

*7.5. Orthogonal Projection

Orthogonal projections are perhaps best introduced with the projection of one vector onto the space generated by another vector. For example, consider two vectors **x** and **y** as drawn in Figure 7.4. As discussed in Chapter 4, the two linearly independent vectors together generate a two-dimensional space, but each vector by itself generates a one-dimensional space. As can be seen in Figure 7.4, a straight line drawn from a point **y** and perpendicular to a vector **x** meets **x** at a point **v**. The vector **v** is said to be the *orthogonal projection* (OP) of **y** on **x** or, more precisely, onto the one-dimensional space generated by **x**. Thus, as this geometric construction of the vector **v** suggests, one can interpret an orthogonal projection as the "*shadow*" of a vector on a given space or subspace.

Suppose, instead of a projection onto a one-dimensional space, we now wish to find the OP of one vector onto a space of two or more dimensions (a space generated by two or more linearly independent vectors). Although the same geometrical argument can be extended to projections onto higher-dimensional spaces, we cannot use geometry to find the OP because we cannot draw such spaces. We have to use numerical methods.

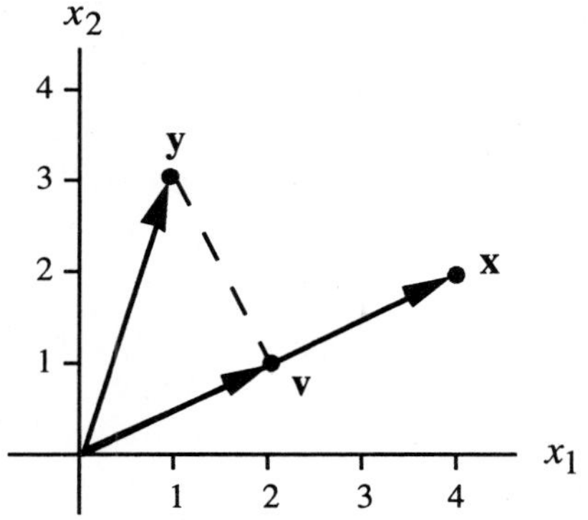

Figure 7.4. The vector **v** is the OP of **y** on the space generated by the vector **x**.

To derive the OP numerically, let us start with the easy case, the OP of one vector onto another vector as above, then generalize to the higher-dimensional case. Referring to Figure 7.4, we first note that the vectors **v** and **x** have the same direction (since we are projecting onto the space of **x**, **v** will always lie in the space of **x**). By Result 4.5(b), we have $\mathbf{v} / \|\mathbf{v}\| = \mathbf{x} / \|\mathbf{x}\|$ and thus it follows that

$$\mathbf{v} = \frac{\mathbf{x}}{\|\mathbf{x}\|}\|\mathbf{v}\|. \tag{7.12}$$

Second, from trigonometry we know that the angle θ between **x** and **y** satisfies

$$\cos\theta = \frac{\|\mathbf{v}\|}{\|\mathbf{y}\|}, \tag{7.13}$$

and from Result 4.1, we also have

$$\cos\theta = \frac{\mathbf{x}^{\mathrm{T}}\mathbf{y}}{\|\mathbf{x}\| \times \|\mathbf{y}\|}. \tag{7.14}$$

From (7.13) and (7.14), we have

$$\frac{\|\mathbf{v}\|}{\|\mathbf{y}\|} = \frac{\mathbf{x}^{\mathrm{T}}\mathbf{y}}{\|\mathbf{x}\| \times \|\mathbf{y}\|},$$

which implies that

$$\|\mathbf{v}\| = \frac{\mathbf{x}^{\mathrm{T}}\mathbf{y}}{\|\mathbf{x}\|}. \tag{7.15}$$

Substituting (7.15) in (7.12), we obtain

$$\mathbf{v} = \frac{\mathbf{x}}{\|\mathbf{x}\|}\frac{\mathbf{x}^T\mathbf{y}}{\|\mathbf{x}\|} = \frac{\mathbf{x}\mathbf{x}^T\mathbf{y}}{\|\mathbf{x}\|^2} = \frac{\mathbf{x}\mathbf{x}^T\mathbf{y}}{\mathbf{x}^T\mathbf{x}}, \tag{7.16}$$

because $\|\mathbf{x}\|^2 = \mathbf{x}^T\mathbf{x}$. Finally we write (7.16) in the equivalent form

$$\mathbf{v} = \mathbf{x}(\mathbf{x}^T\mathbf{x})^{-1}\mathbf{x}^T\mathbf{y} = \mathbf{P}_\mathbf{x}\mathbf{y}, \tag{7.17}$$

where $\mathbf{P}_\mathbf{x} = \mathbf{x}(\mathbf{x}^T\mathbf{x})^{-1}\mathbf{x}^T$ is a square matrix (of what order?). Thus, the projection vector $\mathbf{v}$ of $\mathbf{y}$ on $\mathbf{x}$ can be obtained simply by computing the matrix $\mathbf{P}_\mathbf{x}$ and then pre-multiplying $\mathbf{y}$ by $\mathbf{P}_\mathbf{x}$. What is even more fascinating is the fact that (7.17) applies whether $\mathbf{x}$ is a vector or a matrix.

Suppose now that we wish to project a vector $\mathbf{y}$ onto the space generated by a set of vectors $\mathbf{X}$. (Notice that $\mathbf{X}$ is now a matrix.) Assume for the moment that $\mathbf{X}$ is a full-column rank matrix. We have the following result.

Result 7.2. Let $\mathbf{X}$ be an $n \times k$ matrix of rank k. Then the orthogonal projection of any vector $\mathbf{y}$ of order n onto the space generated by the columns of $\mathbf{X}$ is given by $\mathbf{v} = \mathbf{P}_\mathbf{X}\,\mathbf{y}$, where

$$\mathbf{P}_\mathbf{X} = \mathbf{X}(\mathbf{X}^T\mathbf{X})^{-1}\mathbf{X}^T. \tag{7.18}$$

The matrix $\mathbf{P}_\mathbf{X}$ in (7.18) is clearly a generalization of the matrix $\mathbf{P}_\mathbf{x}$ in (7.17). Thus, the OP of any vector $\mathbf{y}$ of order n onto the space generated by $\mathbf{X}$ is obtained simply by pre-multiplying $\mathbf{y}$ by $\mathbf{P}_\mathbf{X}$. For this reason the matrix $\mathbf{P}_\mathbf{X}$ is called a *projection matrix*. Note also that the orthogonal projection of a vector $\mathbf{y}$ onto the subspace generated by $\mathbf{X}$ is the vector in the subspace that is closest to $\mathbf{y}$. This is because the orthogonal (perpendicular) distance is the shortest distance.

The intuitive notion that the OP of the columns of $\mathbf{X}$ onto the space generated by $\mathbf{X}$ is $\mathbf{X}$ itself follows directly from Result 7.2 because $\mathbf{P}_\mathbf{X}\mathbf{X} = \mathbf{X}(\mathbf{X}^T\mathbf{X})^{-1}\mathbf{X}^T\mathbf{X} = \mathbf{X}\mathbf{I} = \mathbf{X}$, since $(\mathbf{X}^T\mathbf{X})^{-1}\mathbf{X}^T\mathbf{X} = \mathbf{I}$.

Now, let us apply Result 7.2 to the vectors

$$\mathbf{x} = \begin{pmatrix} 4 \\ 2 \end{pmatrix} \quad \text{and} \quad \mathbf{y} = \begin{pmatrix} 1 \\ 3 \end{pmatrix},$$

which are graphed in Figure 7.4. To find the projection of $\mathbf{y}$ on $\mathbf{x}$, we first compute

$$\mathbf{P}_\mathbf{x} = \mathbf{x}(\mathbf{x}^T\mathbf{x})^{-1}\mathbf{x}^T = \begin{pmatrix} 4 \\ 2 \end{pmatrix}\left\{\begin{pmatrix} 4 & 2 \end{pmatrix}\begin{pmatrix} 4 \\ 2 \end{pmatrix}\right\}^{-1}\begin{pmatrix} 4 & 2 \end{pmatrix} = \frac{1}{5}\begin{pmatrix} 4 & 2 \\ 2 & 1 \end{pmatrix}$$

and then obtain the projection vector

$$\mathbf{v} = \mathbf{P}_\mathbf{X}\,\mathbf{y} = \frac{1}{5}\begin{pmatrix} 4 & 2 \\ 2 & 1 \end{pmatrix}\begin{pmatrix} 1 \\ 3 \end{pmatrix} = \begin{pmatrix} 2 \\ 1 \end{pmatrix},$$

which is the same as the vector $\mathbf{v}$ in Figure 7.4.

Let us now look at orthogonal projection in a three-dimensional space. Suppose we wish to project the vector $\mathbf{y}$ onto the space spanned by the vectors $\mathbf{x}_1$ and $\mathbf{x}_2$, where

$$\mathbf{y} = \begin{pmatrix} 4 \\ 3 \\ 4 \end{pmatrix}, \quad \mathbf{x}_1 = \begin{pmatrix} 3 \\ 0 \\ 4 \end{pmatrix}, \quad \text{and} \quad \mathbf{x}_2 = \begin{pmatrix} 4 \\ 0 \\ 2 \end{pmatrix}.$$

We need to form the matrix

$$\mathbf{X} = (\mathbf{x}_1 \quad \mathbf{x}_2) = \begin{pmatrix} 3 & 4 \\ 0 & 0 \\ 4 & 2 \end{pmatrix},$$

and the reader can verify that the corresponding projection matrix is

$$\mathbf{P}_{\mathbf{X}} = \mathbf{X}(\mathbf{X}^{\mathrm{T}}\mathbf{X})^{-1}\mathbf{X}^{\mathrm{T}} = \begin{pmatrix} 1 & 0 & 0 \\ 0 & 0 & 0 \\ 0 & 0 & 1 \end{pmatrix}.$$

Therefore, the OP of $\mathbf{y}$ onto the space spanned by $\mathbf{X}$ is

$$\mathbf{v} = \mathbf{P}_{\mathbf{X}}\mathbf{y} = \begin{pmatrix} 1 & 0 & 0 \\ 0 & 0 & 0 \\ 0 & 0 & 1 \end{pmatrix}\begin{pmatrix} 4 \\ 3 \\ 4 \end{pmatrix} = \begin{pmatrix} 4 \\ 0 \\ 4 \end{pmatrix}.$$

The vectors $\mathbf{y}$, $\mathbf{x}_1$, $\mathbf{x}_2$, and $\mathbf{v}$ are drawn in Figure 7.5, where it is seen that $\mathbf{x}_1$ and $\mathbf{x}_2$ generate a two-dimensional subspace which is given by the (x_1, x_2)-plane and that the OP of $\mathbf{y}$ onto that subspace is the vector $\mathbf{v}$.

As a final example illustrating Result 7.2, and moving toward higher dimensions, suppose we wish to find the OP of the vector $\mathbf{y}$ onto the space generated by the matrix $\mathbf{X}$, where

$$\mathbf{y} = \begin{pmatrix} 122 \\ 115 \\ 128 \\ 145 \\ 108 \end{pmatrix} \quad \text{and} \quad \mathbf{X} = \begin{pmatrix} 10 & 4 \\ 12 & 3 \\ 12 & 4 \\ 14 & 5 \\ 10 & 3 \end{pmatrix}. \tag{7.19}$$

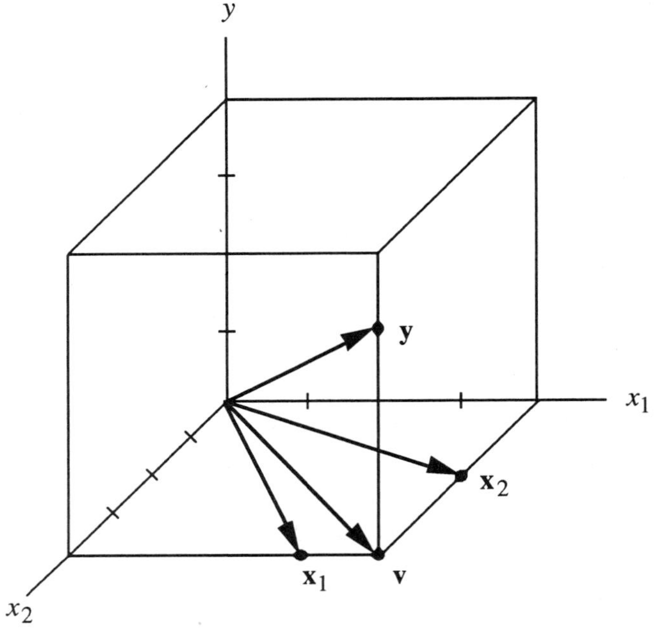

Figure 7.5. The vector $\mathbf{v}$ is the OP of the vector $\mathbf{y}$ onto the space spanned by the vectors $\mathbf{x}_1$ and $\mathbf{x}_2$, in this case the (x_1, x_2)-plane.

We need first to compute

$$\mathbf{X}^{\mathrm{T}}\mathbf{X} = \begin{pmatrix} 10 & 12 & 12 & 14 & 10 \\ 4 & 3 & 4 & 5 & 3 \end{pmatrix} \begin{pmatrix} 10 & 4 \\ 12 & 3 \\ 12 & 4 \\ 14 & 5 \\ 10 & 3 \end{pmatrix} = \begin{pmatrix} 684 & 224 \\ 224 & 75 \end{pmatrix}. \tag{7.20}$$

Since $\det(\mathbf{X}) \neq 0$, $\mathbf{X}^{\mathrm{T}}\mathbf{X}$ is nonsingular and its inverse exists. Using the formula for a matrix inverse given in (6.4), we have (since $\det(\mathbf{X}) = 1124$)

$$(\mathbf{X}^{\mathrm{T}}\mathbf{X})^{-1} = \frac{1}{1124}\begin{pmatrix} 75 & -224 \\ -224 & 684 \end{pmatrix}. \tag{7.21}$$

Hence,

$$\mathbf{P_X} = \mathbf{X}(\mathbf{X}^\mathrm{T}\mathbf{X})^{-1}\mathbf{X}^\mathrm{T}$$

$$= \frac{1}{1124}\begin{pmatrix} 10 & 4 \\ 12 & 3 \\ 12 & 4 \\ 14 & 5 \\ 10 & 3 \end{pmatrix}\begin{pmatrix} 75 & -224 \\ -224 & 684 \end{pmatrix}\begin{pmatrix} 10 & 12 & 12 & 14 & 10 \\ 4 & 3 & 4 & 5 & 3 \end{pmatrix}$$

$$= \frac{1}{281}\begin{pmatrix} 131 & -66 & 58 & 109 & 7 \\ -66 & 207 & 48 & 3 & 93 \\ 58 & 48 & 60 & 74 & 46 \\ 109 & 3 & 74 & 110 & 38 \\ 7 & 93 & 46 & 38 & 54 \end{pmatrix},$$

and the desired projection is

$$\mathbf{v} = \mathbf{P_X y} = \frac{1}{281}\begin{pmatrix} 131 & -66 & 58 & 109 & 7 \\ -66 & 207 & 48 & 3 & 93 \\ 58 & 48 & 60 & 74 & 46 \\ 109 & 3 & 74 & 110 & 38 \\ 7 & 93 & 46 & 38 & 54 \end{pmatrix}\begin{pmatrix} 122 \\ 115 \\ 128 \\ 145 \\ 108 \end{pmatrix} = \begin{pmatrix} 115.22 \\ 115.22 \\ 128.02 \\ 153.63 \\ 102.42 \end{pmatrix}. \qquad (7.22)$$

Therefore, $\mathbf{v}$ is the OP of $\mathbf{y}$ onto the space spanned by the columns of $\mathbf{X}$.

We have assumed so far that $\mathbf{X}$ is of full-column rank so that $(\mathbf{X}^\mathrm{T}\mathbf{X})^{-1}$ exists. If $\mathbf{X}$ is not of full-column rank, we can proceed in either of two ways. First, if $\mathbf{X}$ is rank-deficient, it means that the space generated by its columns can be generated by a full-rank subset of its columns. Therefore, this full-column rank subset can be used in place of $\mathbf{X}$ in Result 7.2, hence there is really no loss of generality to assume that $\mathbf{X}$ is of full-column rank.

Alternatively, we may project onto the column space of $\mathbf{X}$ using a g-inverse of $\mathbf{X}$ (see Section 6.3) as stated in the following result.

Result 7.3. Let $\mathbf{X}$ be an $n \times k$ matrix of rank $r > 0$ and $\mathbf{X}^+$ be the Moore-Penrose inverse of $\mathbf{X}$. Then the projection of any vector $\mathbf{y}$ of order n onto the space generated by the columns of $\mathbf{X}$ is given by

$$\mathbf{v} = \mathbf{Q_X y}, \qquad (7.23)$$

where

$$\mathbf{Q_X} = \mathbf{XX}^+. \qquad (7.24)$$

In fact, Result 7.2 is a special case of Result 7.3. To see this, note that if $\mathbf{X}$ is of full-column rank, by Result 6.11(b), the Moore-Penrose inverse of $\mathbf{X}$ can then be expressed as $\mathbf{X}^+ = (\mathbf{X}^\mathrm{T}\mathbf{X})^{-1}\mathbf{X}^\mathrm{T}$. Hence, $\mathbf{Q_X} = \mathbf{XX}^+ = \mathbf{X}(\mathbf{X}^\mathrm{T}\mathbf{X})^{-1}\mathbf{X}^\mathrm{T} = \mathbf{P_X}$.

Some properties of projection matrices are given in the next two results; other properties are found in Chatterjee and Hadi (1988), Chapter 2.

Result 7.4. For any nonsingular matrix $\mathbf{A}$, the projection matrices for $\mathbf{X}$ and $\mathbf{Z} = \mathbf{XA}$ are the same; that is, $\mathbf{XX}^{+} = \mathbf{ZZ}^{+}$.

Thus, the OP of $\mathbf{y}$ onto the space generated by $\mathbf{X}$ is the same as the OP of $\mathbf{y}$ onto the space generated by $\mathbf{XA}$, for any nonsingular matrix $\mathbf{A}$.

Result 7.5. Let $\mathbf{P}$ be a projection matrix, such as those in (7.18) and (7.24), onto the column space of an $n \times k$ matrix $\mathbf{X}$, with $n \geq k$. Let $\mathbf{I}$ be the identity matrix of order n. Then:

(a) $\mathbf{P}$ is symmetric: $\mathbf{P}^{\mathrm{T}} = \mathbf{P}$.
(b) $\mathbf{P}$ is idempotent: $\mathbf{PP} = \mathbf{P}$.
(c) $\mathrm{tr}(\mathbf{P}) = \mathrm{r}(\mathbf{P}) = \mathrm{r}(\mathbf{X})$.
(d) $\mathbf{I} - \mathbf{P}$ is symmetric and idempotent.
(e) $\mathrm{tr}(\mathbf{I} - \mathbf{P}) = \mathrm{r}(\mathbf{I} - \mathbf{P}) = n - \mathrm{r}(\mathbf{X})$.
(f) $\mathbf{I} - \mathbf{P}$ is a projection matrix onto the orthogonal complement of the columns space of $\mathbf{X}$.

7.6. Singular and Nonsingular Transformations

We have seen in this chapter that a linear transformation involves multiplying a matrix by a vector. Thus, when

$$\mathbf{y} = \mathbf{Ax}, \tag{7.25}$$

we say that $\mathbf{y}$ is a linear transformation of $\mathbf{x}$ or $\mathbf{A}$ linearly transforms $\mathbf{x}$ into $\mathbf{y}$. The effect of the transformation is determined by the transformation matrix $\mathbf{A}$. If $\mathbf{A}$ is square and nonsingular, then the transformation is one-to-one; that is, if we know $\mathbf{x}$ we can uniquely determine $\mathbf{y}$ and, conversely, if we know $\mathbf{y}$ we can uniquely determine $\mathbf{x}$. This two-way transformation is possible because $\mathbf{A}$ is nonsingular; therefore its inverse exists and, multiplying both sides of (7.25) by $\mathbf{A}^{-1}$, we obtain $\mathbf{A}^{-1}\mathbf{y} = \mathbf{A}^{-1}\mathbf{Ax} = \mathbf{x}$. Thus, when $\mathbf{A}$ is nonsingular, we can do the transformation by multiplying $\mathbf{x}$ by $\mathbf{A}$ and undo it by multiplying $\mathbf{y}$ by $\mathbf{A}^{-1}$. Therefore, when $\mathbf{A}$ is nonsingular, both $\mathbf{x}$ and $\mathbf{y}$ lie in the same vector space.

On the other hand, when $\mathbf{A}$ is either singular or not square, the transformation is one-way. That is, if we know $\mathbf{x}$, the relation $\mathbf{y} = \mathbf{Ax}$ determines a unique $\mathbf{y}$, but if we know only $\mathbf{y}$, we cannot uniquely determine $\mathbf{x}$. In other words, there are infinitely many vectors $\mathbf{x}$ satisfying $\mathbf{y} = \mathbf{Ax}$. The reason for this is that when $\mathbf{A}$ is either singular or not square, we can compute a g-inverse of $\mathbf{A}$ and, as we shall see in Section 8.1, $\mathbf{y} = \mathbf{Ax}$ implies $\mathbf{x} = \mathbf{A}^{-}\mathbf{y}$, where $\mathbf{A}^{-}$ is any g-inverse of $\mathbf{A}$; but since there are infinitely many possibilities for $\mathbf{A}^{-}$, we have infinitely many vectors $\mathbf{x}$ satisfying $\mathbf{y} = \mathbf{Ax}$. Note that $\mathbf{y} = \mathbf{Ax}$ implies $\mathbf{x} = \mathbf{A}^{-}\mathbf{y}$ but $\mathbf{x} = \mathbf{A}^{-}\mathbf{y}$ does not imply $\mathbf{y} = \mathbf{Ax}$. Thus, when $\mathbf{A}$ is singular, we cannot undo the transformation, and it is in this sense that singular transformations are one-way.

Note that when $\mathbf{A}$ is not square, the dimension of $\mathbf{y}$ will be different from that of $\mathbf{x}$. Another characteristic of singular transformations is that the vectors $\mathbf{x}$ and $\mathbf{y}$ lie in different vector spaces. To see this, note that, when $\mathbf{A}$ is singular, $\mathbf{y}$ can be written as

$$\mathbf{y} = \mathbf{A}\mathbf{x} = x_1\,\mathbf{a}_1 + x_2\,\mathbf{a}_2 + \cdots + x_n\,\mathbf{a}_n,$$

where $x_1, x_2, ..., x_n$ are the elements of $\mathbf{x}$ and $\boldsymbol{a}_1, \boldsymbol{a}_2, ..., \boldsymbol{a}_n$ are the n columns of $\mathbf{A}$. But the singularity of $\mathbf{A}$ implies that $\boldsymbol{a}_1, \boldsymbol{a}_2, ..., \boldsymbol{a}_n$ are linearly dependent. Hence $\mathbf{y}$ can be written as a linear combination of fewer than n vectors, which implies that $\mathbf{y}$ lies in a smaller subspace, that is, a subspace smaller than $\Re^n$.

One can then classify linear transformations as singular or nonsingular depending on whether the transformation matrix is singular or nonsingular. Of the transformations we have discussed earlier in this chapter, orthogonal transformations are nonsingular transformations because all orthogonal matrices are nonsingular; oblique transformations can be either singular or nonsingular; and orthogonal projections are always singular unless the projection matrix is the identity matrix. The reason for this is that, except for the identity matrix, all orthogonal projection matrices are idempotent, hence singular (see Exercise 7.8). (But if the matrix of transformation is the identity matrix, we would have $\mathbf{y} = \mathbf{I}\mathbf{x} = \mathbf{x}$ and in effect no transformation has occurred.)

Exercises

7.1. Let $\mathbf{x} = (x_1\;\; x_2)^{\mathrm{T}}$ and $\mathbf{y} = (y_1\;\; y_2)^{\mathrm{T}}$. Find $\mathbf{A}$ and $\mathbf{c}$ so that the two equations in each of the following cases can be written as $\mathbf{y} = \mathbf{A}(\mathbf{x} - \mathbf{c})$:
(a) $y_1 = x_1 + 2x_2 - 1$ and $y_2 = 2x_1 + x_2 - 2$.
(b) $\sqrt{2}\,y_1 = x_1 - x_2 - 1$ and $\sqrt{2}\,y_2 = x_1 + x_2 - 1$.
(c) $y_1 = 2x_1 + 2$ and $y_2 = 4x_2 - 4$.
(d) $y_1 = 2x_2 - 4$ and $y_2 = 4x_1 - 4$.

7.2. Given $\mathbf{A} = \dfrac{1}{\sqrt{2}}\begin{pmatrix} 1 & 1 \\ -1 & 1 \end{pmatrix}$, $\mathbf{v}_1 = \begin{pmatrix} 1 \\ 1 \end{pmatrix}$, and $\mathbf{v}_2 = \begin{pmatrix} 0 \\ 4 \end{pmatrix}$:
(a) Compute $\mathbf{w}_1 = \mathbf{A}\mathbf{v}_1$ and $\mathbf{w}_2 = \mathbf{A}\mathbf{v}_2$.
(b) Compute the angle between $\mathbf{w}_1$ and the horizontal axis and the angle between $\mathbf{w}_2$ and the horizontal axis.
(c) Graph the vectors $\mathbf{v}_1$ and $\mathbf{v}_2$ with respect to the conventional axes.
(d) On the same graph draw the new set of axes (as in Figure 7.1) for which the elements of the vectors $\mathbf{w}_1$ and $\mathbf{w}_2$ can be regarded as the coordinates of the vectors $\mathbf{v}_1$ and $\mathbf{v}_2$ with respect to the new set of axes.

7.3. Repeat Exercise 7.2 with $\mathbf{A} = \dfrac{1}{\sqrt{2}}\begin{pmatrix} 1 & -1 \\ 1 & 1 \end{pmatrix}$.

7.4. Repeat Exercise 7.2 with $\mathbf{A} = \begin{pmatrix} 0 & 1 \\ 1 & 0 \end{pmatrix}$.

7.5. Find the 2×2 matrix $\mathbf{A}$ that can be used to rotate the original set of axes:
(a) counterclockwise by an angle of 60°. (b) clockwise by an angle of 60°.
(c) counterclockwise by an angle of 90°. (d) clockwise by an angle of 90°.
(e) clockwise by an angle of 180°. (f) clockwise by an angle of 360°.

7.6. Graph the vectors

$$\mathbf{a} = \begin{pmatrix} 0 \\ 1 \end{pmatrix}, \quad \mathbf{b} = \begin{pmatrix} -1 \\ 1 \end{pmatrix}, \quad \mathbf{c} = \begin{pmatrix} 1 \\ 0 \end{pmatrix}, \quad \mathbf{d} = \begin{pmatrix} 2 \\ -2 \end{pmatrix},$$

as well as the vectors obtained by the following orthogonal projections:
(a) **a** onto **b** (b) **a** onto **c** (c) **a** onto **d**
(d) **b** onto **d** (e) **b** onto the x-axis (f) **b** onto the y-axis

7.7. Consider the following vectors:

$$\mathbf{a} = \begin{pmatrix} 0 \\ 1 \\ 2 \end{pmatrix}, \quad \mathbf{b} = \begin{pmatrix} -1 \\ 1 \\ -1 \end{pmatrix}, \quad \mathbf{c} = \begin{pmatrix} 1 \\ 0 \\ 3 \end{pmatrix}, \quad \mathbf{d} = \begin{pmatrix} 2 \\ -2 \\ 2 \end{pmatrix}.$$

Compute (and comment on the results of) the projection of the vector **a** onto the space spanned by:
(a) the vector **b**. (b) the vector **c**. (c) the vector **d**.
(d) the matrix (**b** : **c**). (e) the matrix (**b** : **d**). (f) the matrix (**c** : **d**).
(g) the matrix (**a** : **b**). (h) the matrix (**b** : **a**). (i) the matrix (**b** : **c** : **d**).

7.8. Prove that the only nonsingular idempotent matrix is the identity matrix, hence an orthogonal projection is an example of a singular linear transformation.

7.9. Suppose that just after the transformation of a vector **x** to a vector $\mathbf{y} = \mathbf{Ax}$ using a computer, the vector **x** was accidentally deleted. Is it possible to recover the vector **x** in each of the following cases? If yes, compute **x**, and if no, indicate why not.

(a) $\mathbf{A} = \begin{pmatrix} 1 & 0 \\ 0 & 2 \end{pmatrix}$ and $\mathbf{y} = \begin{pmatrix} 1 \\ 2 \end{pmatrix}$. (b) $\mathbf{A} = \begin{pmatrix} 1 & -1 \\ 1 & 2 \end{pmatrix}$ and $\mathbf{y} = \begin{pmatrix} 0 \\ 0 \end{pmatrix}$.

(c) $\mathbf{A} = \begin{pmatrix} 1 & 2 \\ 2 & 4 \end{pmatrix}$ and $\mathbf{y} = \begin{pmatrix} 1 \\ 1 \end{pmatrix}$. (d) $\mathbf{A} = \begin{pmatrix} 0 & 1 \\ 1 & 0 \end{pmatrix}$ and $\mathbf{y} = \begin{pmatrix} 2 \\ 3 \end{pmatrix}$.

(e) $\mathbf{A} = \frac{1}{2}\begin{pmatrix} 1 & 1 \\ -1 & 1 \end{pmatrix}$ and $\mathbf{y} = \begin{pmatrix} 2 \\ 1 \end{pmatrix}$. (f) $\mathbf{A} = \frac{1}{\sqrt{2}}\begin{pmatrix} 1 & 1 \\ -1 & 1 \end{pmatrix}$ and $\mathbf{y} = \begin{pmatrix} 2 \\ 1 \end{pmatrix}$.

7.10. Determine whether each of the transformations in Exercise 7.9 is:
(a) a nonsingular transformation. (b) a singular transformation.
(c) an orthogonal transformation. (d) an oblique transformation.
(e) an orthogonal projection.

8 Some Applications

Matrix algebra has numerous applications in many fields of study. In this chapter we illustrate the following applications of the matrix algebra tools we have learned so far: (a) solving systems of simultaneous linear equations, (b) simplifying some descriptive statistics, (c) formulating linear regression models, and (d) updating computations. Other examples of matrix applications are given in Chapters 10 and 11.

8.1. Simultaneous Linear Equations

8.1.1. Definitions

Suppose we wish to find two numbers x_1 and x_2 whose sum is equal to 5 and whose difference is equal to 1. This can be translated into two equations, one for the sum and one for the difference, as follows:

$$\begin{aligned} x_1 + x_2 &= 5 \\ x_1 - x_2 &= 1 \end{aligned} \tag{8.1}$$

These two equations can equivalently be rewritten as

$$\begin{aligned} x_2 &= 5 - x_1 \\ x_2 &= -1 + x_1 \end{aligned} \tag{8.2}$$

Each of these two equations represents a straight-line relationship between x_1 and x_2: The first is an equation of the straight line with a slope of -1 and an x_2-intercept of 5; the second is an equation of the straight line with a slope of 1 and an x_2-intercept of -1. Because both equations represent straight lines, (8.1) is referred to as a *system of linear equations*. This system consists of two linear equations with two unknowns x_1 and x_2. The objective is to solve (8.1) for the unknowns, that is, to find values for x_1 and x_2 that satisfy the two equations simultaneously. For this reason the system is also called a system of *simultaneous* linear equations.

The system in (8.1) can be extended to the more general case where there are m equations and n unknowns. A general system of linear equations can be written as

$$a_{11}\,x_1 + a_{12}\,x_2 + \cdots + a_{1n}\,x_n = b_1$$

$$a_{21}\,x_1 + a_{22}\,x_2 + \cdots + a_{2n}\,x_n = b_2$$

$$\vdots$$

$$a_{m1}\,x_1 + a_{m2}\,x_2 + \cdots + a_{mn}\,x_n = b_m$$

where $x_1, x_2, \ldots, x_n$ are unknown variables, a_{ij} are known coefficients, and b_1, $b_2, \ldots, b_m$ are known constants.

The above system can be simplified considerably using matrix notation. Create the matrices

$$\mathbf{A} = \begin{pmatrix} a_{11} & a_{12} & \cdots & a_{1n} \\ a_{21} & a_{22} & \cdots & a_{2n} \\ \vdots & \vdots & \ddots & \vdots \\ a_{m1} & a_{m2} & \cdots & a_{mn} \end{pmatrix}, \quad \mathbf{x} = \begin{pmatrix} x_1 \\ x_2 \\ \vdots \\ x_n \end{pmatrix}, \quad \mathbf{b} = \begin{pmatrix} b_1 \\ b_2 \\ \vdots \\ b_m \end{pmatrix}$$

and verify that the m equations above can be written in matrix notation as

$$\mathbf{Ax} = \mathbf{b}. \tag{8.3}$$

For example, the first equation is obtained by multiplying the first row of **A** by the vector **x**, yielding b_1. Note here that the notation in (8.3) is independent of both m and n; that is, regardless of the number of equations and the number of unknowns in a system of linear equations, the system can always be expressed as in (8.3). The matrix **A** is called the *coefficient matrix*, the vector **x** is the *vector of unknowns*, and the vector **b** is called the *right-hand side* or the *vector of constants*.

For our particular example in (8.1), we have

$$\mathbf{A} = \begin{pmatrix} 1 & 1 \\ 1 & -1 \end{pmatrix}, \quad \mathbf{x} = \begin{pmatrix} x_1 \\ x_2 \end{pmatrix}, \quad \mathbf{b} = \begin{pmatrix} 5 \\ 1 \end{pmatrix} \tag{8.4}$$

so that when we substitute these matrices in (8.3) and multiply, we obtain the two equations in (8.1).

8.1.2. Graphical Solution

When $n = 2$, as is the case of (8.1), the system can also be solved graphically. The two lines in (8.1) are drawn in Figure 8.1. They intersect at $(x_1, x_2) = (3, 2)$. Since this point is on both lines, its coordinates satisfy the two equations in (8.1) and give the unique solution of the system. Note that if the two lines were parallel, there would be no solution; and if they were identical, there would be infinitely many solutions.

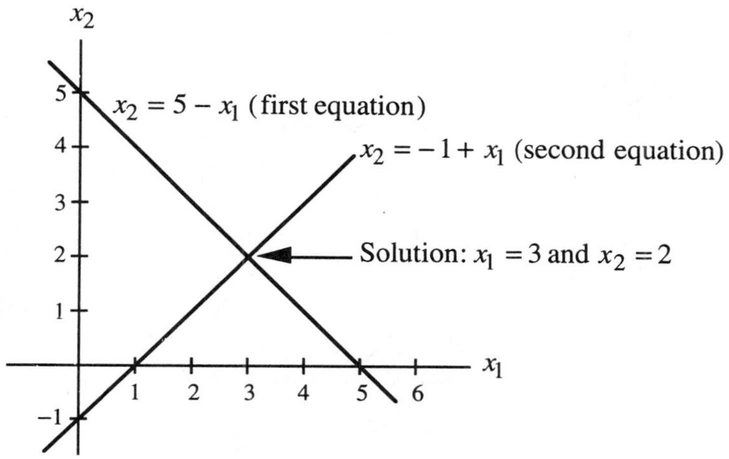

Figure 8.1. Graphical solution of the linear system in (8.1).

The graph in Figure 8.1 may help us explain all possible situations that may arise when one tries to solve a system of linear equations. To illustrate, suppose we wish to solve the following system of linear equations:

$$\begin{aligned} x_1 + x_2 &= 5 \\ x_1 - x_2 &= 1 \\ 2x_1 + 2x_2 &= 10 \end{aligned} \tag{8.5}$$

Notice that the first two equations are the same as in (8.1). Dividing both sides of the third equation by 2, we obtain the first equation, and so every pair (x_1, x_2) that satisfies the first equation will also satisfy the third. This means that every point that lies on the line represented by the first equation also lies on the line represented by the third. Since the first and third equations are represented by the same line in Figure 8.1, the solution to (8.5) is the same as that of (8.1). This solution is still unique. The reason for this is that the third equation is in a sense redundant because it did not add any new information to the system of the two equations. When solving systems of linear equations, it is customary to delete all redundant equations from the system.

Let us now try to solve a third system of linear equations:

$$\begin{aligned} x_1 + x_2 &= 5 \\ x_1 - x_2 &= 1 \\ x_1 - 4x_2 &= 0 \end{aligned} \tag{8.6}$$

This system differs from that in (8.5) only in the third equation. The third equation can be written as $x_2 = 0.25x_1$, which represents a straight line with a slope of 0.25 and an x_2-intercept of 0. The three lines are drawn in Figure 8.2.

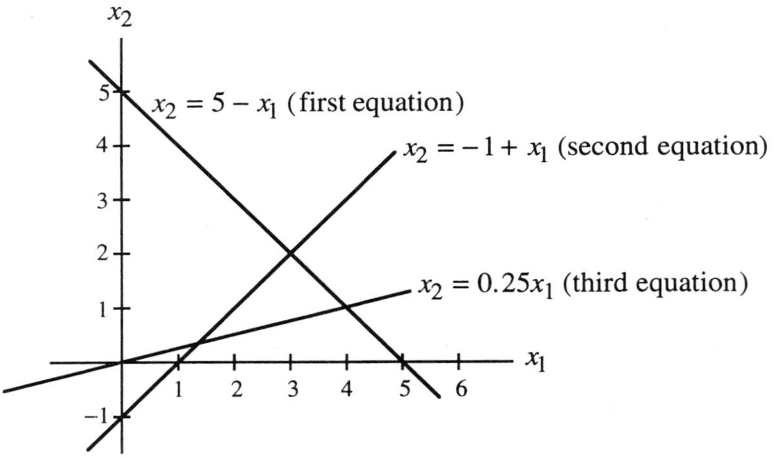

Figure 8.2. Graphical representation of the linear system in (8.6).

Notice that there are three intersection points, each of which satisfies only two of the three equations, but none of which satisfies all three equations. We therefore conclude that the system in (8.6) has no solution. Such a system is said to be *inconsistent* (here, any one equation is inconsistent with the other two).

So far we have seen two types of linear systems; one has a unique solution and the other has no solution at all. There is a third and final possibility. To illustrate, suppose we wish to solve the following system of linear equations:

$$\begin{aligned} x_1 + x_2 &= 5 \\ 3x_1 + 3x_2 &= 15 \end{aligned} \tag{8.7}$$

Note that one of these two equations is redundant (the second can be obtained by multiplying the first by 3). If we delete one equation, we will end up with only one equation in two unknowns. We can then set either unknown to any arbitrary value and solve for the other, and so produce infinitely many solutions. Note that if we graphed the two lines they would be identical, and any point on the line would be a solution to (8.7) because any such point would satisfy the two equations. A linear system for which infinitely many solutions exist is said to be *indeterminate*.

8.1.3. Numerical Solution

A graphical solution to a system of linear equations can be obtained when we have only two unknowns, that is, when $n = 2$. When $n > 2$, we have to rely on numerical methods. This section describes numerical methods for solving systems of linear equations.

The graphical solution presented above indicates that when we try to solve a system of linear equations, we will end up with one and only one of three possible situations: a unique solution, no solution, or infinitely many solutions. A linear system that has at least one solution is said to be *consistent*. Note that if the system has at least one solution, (8.3) implies that $\mathbf{b}$ can be written as a linear combination of the columns of $\mathbf{A}$. Therefore, if we augment the matrix $\mathbf{A}$ by the vector $\mathbf{b}$, the columns of the augmented matrix

$$\mathbf{B} = (\mathbf{A} : \mathbf{b}) \tag{8.8}$$

will be linearly dependent, in which case $\mathbf{B}$ will have the same rank as $\mathbf{A}$. Since this is true for any system that has a solution, it follows that if $\text{r}(\mathbf{B}) > \text{r}(\mathbf{A})$, the system of linear equations has no solution and the system is said to be *inconsistent*.

On the other hand, if $\text{r}(\mathbf{B}) = \text{r}(\mathbf{A})$, then the system has at least one solution. If $\text{r}(\mathbf{B}) = \text{r}(\mathbf{A}) = n$, where n is the number of unknowns, the system has a unique solution; otherwise the system has infinitely many solutions. When the system has infinitely many solutions, one such solution can always be obtained by setting $n - \text{r}(\mathbf{A})$ of the variables to arbitrary values and solving for the remaining variables. The variables which can be set to arbitrary values are called *free* variables.

The above discussion is summarized as follows (see also Figure 8.3).

Result 8.1. For the system of linear equations defined in (8.3), let $\mathbf{B}$ be the augmented matrix of (8.8); then exactly one of the following must hold:

(a) Inconsistent system: the system has no solution if $\text{r}(\mathbf{B}) > \text{r}(\mathbf{A})$.

(b) Unique solution: the system has a unique solution if $\text{r}(\mathbf{A}) = n$.

(c) Many solutions: the system has infinitely many solutions if $\text{r}(\mathbf{B}) = \text{r}(\mathbf{A}) < n$. A solution can be obtained by setting $n - \text{r}(\mathbf{A})$ of the variables arbitrarily, then solving for the remaining variables uniquely.

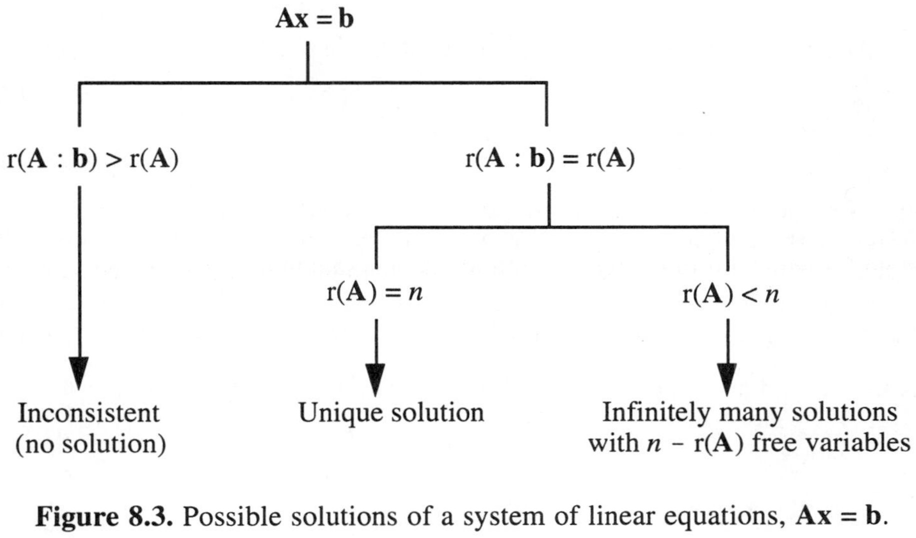

Figure 8.3. Possible solutions of a system of linear equations, $\mathbf{Ax} = \mathbf{b}$.

We shall see shortly that, in cases where there are infinitely many solutions, these solutions can be generated more systematically using a generalized inverse (see Section 6.3). A simple numerical algorithm for solving systems of linear equations is given in the appendix to this chapter. We conclude this section with two special cases of (8.3). The first is the case where **A** is square, that is, the number of equations is equal to the number of unknowns, and the second case is when $\mathbf{b} = \mathbf{0}$.

8.1.4. The Case of Square Coefficient Matrix

When the coefficient matrix **A** is square, we have two subcases: **A** is either nonsingular or singular. When **A** is nonsingular, we can compute $\mathbf{A}^{-1}$, then pre-multiply both sides of (8.3) by $\mathbf{A}^{-1}$ and obtain

$$\mathbf{A}^{-1}\mathbf{A}\mathbf{x} = \mathbf{A}^{-1}\mathbf{b}.$$

But, since $\mathbf{A}^{-1}\mathbf{A} = \mathbf{I}$ and $\mathbf{I}\mathbf{x} = \mathbf{x}$, it follows that

$$\mathbf{x} = \mathbf{A}^{-1}\mathbf{b} \tag{8.9}$$

is the unique solution to (8.3). Thus, if **A** is square and nonsingular, the solution of the system is obtained directly by pre-multiplying **b** by $\mathbf{A}^{-1}$. Let us now apply (8.9) to the linear system in (8.2), for which the corresponding matrix **A** and vector **b** are

$$\mathbf{A} = \begin{pmatrix} 1 & 1 \\ 1 & -1 \end{pmatrix} \quad \text{and} \quad \mathbf{b} = \begin{pmatrix} 5 \\ 1 \end{pmatrix},$$

as given in (8.4). Using (6.4) to compute $\mathbf{A}^{-1}$, we have

$$\mathbf{A}^{-1} = \frac{1}{-2}\begin{pmatrix} -1 & -1 \\ -1 & 1 \end{pmatrix}.$$

Thus, the solution of the system of linear equations in (8.2) is given by

$$\mathbf{x} = \mathbf{A}^{-1}\mathbf{b} = \frac{1}{-2}\begin{pmatrix} -1 & -1 \\ -1 & 1 \end{pmatrix}\begin{pmatrix} 5 \\ 1 \end{pmatrix} = \begin{pmatrix} 3 \\ 2 \end{pmatrix}. \tag{8.10}$$

Thus, $x_1 = 3$ and $x_2 = 2$ is the unique solution to the system in (8.2).

The second subcase occurs when **A** is singular, so we cannot use (8.9) because $\mathbf{A}^{-1}$ does not exist. In this case, if $\mathrm{r}(\mathbf{A}) < \mathrm{r}(\mathbf{B})$, then the system is inconsistent, and if $\mathrm{r}(\mathbf{A}) = \mathrm{r}(\mathbf{B})$, then the system has infinitely many solutions. These solutions can be generated using a g-inverse of **A** (see Section 6.3) as stated in the following result.

Result 8.2. Suppose the system of linear equations defined in (8.3) is consistent and let $\mathbf{A}^-$ be any g-inverse of **A**; then:
(a) A solution can be obtained by

$$\mathbf{x} = \mathbf{A}^-\mathbf{b}. \tag{8.11}$$

(b) All other solutions can be expressed as

$$\mathbf{x} = \mathbf{A}^{-}\mathbf{b} + (\mathbf{I} - \mathbf{A}^{-}\mathbf{A})\mathbf{c}, \tag{8.12}$$

where $\mathbf{c}$ is any arbitrary vector.

Note that Result 8.2 assumes that the system is consistent, that is, it has at least one solution. In this case, the following remarks are noteworthy:

1. When $\mathbf{A}$ is nonsingular, $\mathbf{A}^{-} = \mathbf{A}^{-1}$, and both (8.11) and (8.12) reduce to (8.9).
2. The statement of Result 8.2 does not require $\mathbf{A}$ to be square; so as long as the system is consistent, a solution can always be found using (8.11), and other solutions can be obtained using (8.12).
3. If $r(\mathbf{A}) = n$ and we choose $\mathbf{A}^{-}$ to be $\mathbf{A}^{+}$, the Moore-Penrose inverse of $\mathbf{A}$, then by Result 6.11(b), $\mathbf{A}^{+}\mathbf{A} = \mathbf{I}$, hence (8.12) reduces to (8.11), which means that we have a unique solution and (8.11) can be obtained directly by pre-multiplying both sides of (8.3) by $\mathbf{A}^{+}$.

8.1.5. Homogeneous Systems

The second special case of a linear system occurs when $\mathbf{b} = \mathbf{0}$. In this case, (8.3) becomes

$$\mathbf{A}\mathbf{x} = \mathbf{0} \tag{8.13}$$

and the system is called a *homogeneous* system. The augmented matrix here is $\mathbf{B} = (\mathbf{A} : \mathbf{0})$, which implies that $r(\mathbf{B}) = r(\mathbf{A})$, which in turn implies that homogeneous systems are always *consistent*, that is, they have at least one solution. This is clearly the case, because in any homogeneous system there always exists the solution $\mathbf{x} = \mathbf{0}$. The solution $\mathbf{x} = \mathbf{0}$ is called the *trivial* solution. If $\mathbf{A}$ is nonsingular, the trivial solution is the only solution; otherwise there are infinitely many solutions to a homogeneous system. These solutions can be generated by (8.12).

8.2. Some Descriptive Statistics

In this section we show how matrix notation can be used to simplify the formulas of some descriptive statistics such as the *mean*, *variance*, *covariance*, and *correlation coefficient*. Suppose we draw a (preferably random) sample of n individuals from a population of interest. For each individual, we measure k characteristics or variables, such as age, height, weight, and so on. The data can be organized in an $n \times k$ matrix $\mathbf{X}$ as follows:

$$\mathbf{X} = \begin{pmatrix} x_{11} & x_{12} & \cdots & x_{1k} \\ x_{21} & x_{22} & \cdots & x_{2k} \\ \vdots & \vdots & \ddots & \vdots \\ x_{n1} & x_{n2} & \cdots & x_{nk} \end{pmatrix}.$$

Here the ith row of $\mathbf{X}$ represents k measurements on the ith individual, and the jth column represents the jth variable, namely, measurements of a single characteristic for all individuals.

For example, the heights (in centimeters) and weights (in kilograms) of a sample of five individuals are organized in the matrix

$$\mathbf{X} = \begin{pmatrix} 178 & 76 \\ 180 & 80 \\ 170 & 70 \\ 172 & 74 \\ 175 & 75 \end{pmatrix}. \tag{8.14}$$

The first column represents the variable height and the second represents the variable weight. The sample is chosen here to be small so that the reader could easily do the calculations by hand. Here $n = 5$ and $k = 2$.

The goal is to compute some descriptive statistics for the data in $\mathbf{X}$, such as means, variances, covariances, and correlation coefficients for the k variables in $\mathbf{X}$.

8.2.1. The Mean Vector

Let $\mathbf{x}_j$ be the jth variable (column of $\mathbf{X}$). The mean of $\mathbf{x}_j$ is defined by

$$\bar{x}_j = n^{-1} \sum_{i=1}^{n} x_{ij}\,, \quad j = 1, \ldots, k.$$

($\bar{x}$ is read as x-bar.) For example, using $\mathbf{X}$ in (8.14), we have

$$\bar{x}_1 = \frac{178 + 180 + 170 + 172 + 175}{5} = 175$$

and

$$\bar{x}_2 = \frac{76 + 80 + 70 + 74 + 75}{5} = 75,$$

Thus, the mean height is 175 centimeters and the mean weight is 75 kilograms. These two means can be arranged in a vector:

$$\bar{\mathbf{x}} = \begin{pmatrix} 175 \\ 75 \end{pmatrix}. \tag{8.15}$$

8.2.2. The Variance-Covariance Matrix

The covariance between any two variables $\mathbf{x}_j$ and $\mathbf{x}_l$ is defined by

$$s_{jl} = (n-1)^{-1} \sum_{i=1}^{n} (x_{ij} - \bar{x}_j)(x_{il} - \bar{x}_l), \; j = 1, \ldots, k \text{ and } l = 1, \ldots, k. \qquad (8.16)$$

Note that in (8.16), when $j = l$, s_{jl} becomes

$$s_{jj} = (n-1)^{-1} \sum_{i=1}^{n} (x_{ij} - \bar{x}_j)^2 = s_j^2, \; j = 1, \ldots, k, \qquad (8.17)$$

which is the variance of $\mathbf{x}_j$.[1] Thus, as is well known in statistics, the covariance between a variable and itself is simply the variance of the variable. The positive square root of the variance is known as the *standard deviation*.

For example, using the matrix $\mathbf{X}$ in (8.14), we obtain

$$\begin{aligned} s_{11} &= \frac{(178-175)^2 + (180-175)^2 + (170-175)^2 + (172-175)^2 + (175-175)^2}{5-1} \\ &= 17, \end{aligned}$$

$$s_{22} = \frac{(76-75)^2 + (80-75)^2 + (70-75)^2 + (74-75)^2 + (75-75)^2}{5-1} = 13,$$

$$s_{12} = \frac{(178-175)(76-75) + (180-175)(80-75) + \cdots + (175-175)(75-75)}{5-1} = 14.$$

It also follows from (8.16) that $s_{21} = s_{12}$. Thus, the variance of the variable height is 17, the variance of the variable weight is 13, and the covariance between the two variables is 14. These variances and covariances can be arranged in a matrix:

$$\mathbf{S} = \begin{pmatrix} s_{11} & s_{12} \\ s_{21} & s_{22} \end{pmatrix} = \begin{pmatrix} 17 & 14 \\ 14 & 13 \end{pmatrix}. \qquad (8.18)$$

In general, we can arrange the means in a vector $\bar{\mathbf{x}}$, and the variances and covariances in a matrix $\mathbf{S}$ as follows:

$$\bar{\mathbf{x}} = \begin{pmatrix} \bar{x}_1 \\ \bar{x}_2 \\ \vdots \\ \bar{x}_k \end{pmatrix} \quad \text{and} \quad \mathbf{S} = \begin{pmatrix} s_{11} & s_{12} & \cdots & s_{1k} \\ s_{21} & s_{22} & \cdots & s_{2k} \\ \vdots & \vdots & \ddots & \vdots \\ s_{k1} & s_{k2} & \cdots & s_{kk} \end{pmatrix}. \qquad (8.19)$$

The vector $\bar{\mathbf{x}}$ is referred to as the *mean* vector, and the matrix $\mathbf{S}$ is referred to as the *variance-covariance matrix*. The diagonal elements of $\mathbf{S}$ are the variances, and the

[1] Actually, if the data represent a sample from a given population, the means, variances, and covariances as defined here are the estimates of the population means, variances, and covariances, respectively.

off-diagonal elements are the covariances. From (8.16) it follows that $s_{lj} = s_{jl}$ (the covariance between $\mathbf{x}_l$ and $\mathbf{x}_j$ is the same as the covariance between $\mathbf{x}_j$ and $\mathbf{x}_l$). Thus, $\mathbf{S}$ is a symmetric matrix, i.e., $\mathbf{S}^{\mathrm{T}} = \mathbf{S}$.

Using matrix notation we can write $\bar{\mathbf{x}}$ and $\mathbf{S}$ as follows. Let $\mathbf{1}$ be a vector of n ones, then compute

$$n^{-1}\,\mathbf{X}^{\mathrm{T}}\mathbf{1} = n^{-1}\begin{pmatrix} x_{11} & x_{12} & \cdots & x_{1k} \\ x_{21} & x_{22} & \cdots & x_{2k} \\ \vdots & \vdots & \ddots & \vdots \\ x_{n1} & x_{n2} & \cdots & x_{nk} \end{pmatrix}\begin{pmatrix} 1 \\ 1 \\ \vdots \\ 1 \end{pmatrix}$$

$$= n^{-1}\begin{pmatrix} \sum_{i=1}^{n} x_{i1} \\ \sum_{i=1}^{n} x_{i2} \\ \vdots \\ \sum_{i=1}^{n} x_{ik} \end{pmatrix} = \begin{pmatrix} \bar{x}_1 \\ \bar{x}_2 \\ \vdots \\ \bar{x}_k \end{pmatrix} = \bar{\mathbf{x}}. \tag{8.20}$$

Thus, the mean vector $\bar{\mathbf{x}}$ in (8.19) can simply be written as

$$\bar{\mathbf{x}} = n^{-1}\,\mathbf{X}^{\mathrm{T}}\mathbf{1}. \tag{8.21}$$

To illustrate, using the matrix $\mathbf{X}$ in (8.14), we compute

$$\bar{\mathbf{x}} = \frac{1}{5}\begin{pmatrix} 178 & 180 & 170 & 172 & 175 \\ 76 & 80 & 70 & 74 & 75 \end{pmatrix}\begin{pmatrix} 1 \\ 1 \\ 1 \\ 1 \\ 1 \end{pmatrix} = \begin{pmatrix} 175 \\ 75 \end{pmatrix},$$

which is the same as $\bar{\mathbf{x}}$ in (8.15).

To compute the variance-covariance matrix, we need first to subtract the mean of the jth variable, $\bar{x}_j$, from each value of x_{ij}. This is called *centering* and the resultant matrix is called the *centered* matrix. The centered matrix, denoted by $\tilde{\mathbf{X}}$, is

$$\tilde{\mathbf{X}} = \begin{pmatrix} x_{11} & x_{12} & \cdots & x_{1k} \\ x_{21} & x_{22} & \cdots & x_{2k} \\ \vdots & \vdots & \ddots & \vdots \\ x_{n1} & x_{n2} & \cdots & x_{nk} \end{pmatrix} - \begin{pmatrix} \bar{x}_1 & \bar{x}_2 & \cdots & \bar{x}_k \\ \bar{x}_1 & \bar{x}_2 & \cdots & \bar{x}_k \\ \vdots & \vdots & \ddots & \vdots \\ \bar{x}_1 & \bar{x}_2 & \cdots & \bar{x}_k \end{pmatrix}. \tag{8.22}$$

Notice that the first of the two matrices on the right is **X** and that all the rows of the second matrix are equal to $\bar{\mathbf{x}}^{\mathrm{T}}$, the transpose of the mean vector. For example, the centered matrix corresponding to **X** in (8.14) is

$$\tilde{\mathbf{X}} = \begin{pmatrix} 178 & 76 \\ 180 & 80 \\ 170 & 70 \\ 172 & 74 \\ 175 & 75 \end{pmatrix} - \begin{pmatrix} 175 & 75 \\ 175 & 75 \\ 175 & 75 \\ 175 & 75 \\ 175 & 75 \end{pmatrix} = \begin{pmatrix} 3 & 1 \\ 5 & 5 \\ -5 & -5 \\ -3 & -1 \\ 0 & 0 \end{pmatrix}. \tag{8.23}$$

Notice that the sum of the elements in each column of $\tilde{\mathbf{X}}$ is zero. This is always true, because the elements in the jth column of any matrix **X** satisfy

$$\sum_{i=1}^{n}(x_{ij} - \bar{x}_j) = 0. \tag{8.24}$$

The proof of this result is simple and is left as an exercise for the reader.

The second matrix on the right-hand side of (8.22) can be expressed as

$$\mathbf{1}\bar{\mathbf{x}}^{\mathrm{T}} = \begin{pmatrix} 1 \\ 1 \\ \vdots \\ 1 \end{pmatrix} \begin{pmatrix} \bar{x}_1 & \bar{x}_2 & \cdots & \bar{x}_k \end{pmatrix} = \begin{pmatrix} \bar{x}_1 & \bar{x}_2 & \cdots & \bar{x}_k \\ \bar{x}_1 & \bar{x}_2 & \cdots & \bar{x}_k \\ \vdots & \vdots & \ddots & \vdots \\ \bar{x}_1 & \bar{x}_2 & \cdots & \bar{x}_k \end{pmatrix}.$$

Using (8.21) and Result 2.2(a) and (d), $\bar{\mathbf{x}}^{\mathrm{T}} = n^{-1}\, \mathbf{1}^{\mathrm{T}}\mathbf{X}$, hence $\mathbf{1}\bar{\mathbf{x}}^{\mathrm{T}} = n^{-1}\, \mathbf{1}\mathbf{1}^{\mathrm{T}}\mathbf{X}$. (Note that since n^{-1} is a scalar, it can be placed before or after a matrix.) We can therefore write $\tilde{\mathbf{X}}$ as

$$\begin{aligned} \tilde{\mathbf{X}} &= \mathbf{X} - n^{-1}\, \mathbf{1}\mathbf{1}^{\mathrm{T}}\mathbf{X} \\ &= (\mathbf{I} - n^{-1}\, \mathbf{1}\mathbf{1}^{\mathrm{T}})\mathbf{X} = \mathbf{C}\mathbf{X}, \end{aligned} \tag{8.25}$$

where

$$\mathbf{C} = (\mathbf{I} - n^{-1}\, \mathbf{1}\mathbf{1}^{\mathrm{T}}). \tag{8.26}$$

Note that $\mathbf{1}\mathbf{1}^{\mathrm{T}}$ is simply an $n \times n$ matrix of ones. Note also that when **X** is factored out to the right, it is replaced by the identity matrix so that **I** and $n^{-1}\, \mathbf{1}\mathbf{1}^{\mathrm{T}}$ are conformable for subtraction. The $n \times n$ matrix **C** is called the *centering* matrix because when we pre-multiply **X** by **C** we obtain $\tilde{\mathbf{X}}$, which is the centered **X** matrix. Thus, (8.25) is an alternative way to compute the centered matrix in (8.22).

As an example, let us compute the centering matrix required to center the matrix **X** in (8.14). Here $n = 5$, so we have

$$\mathbf{C} = (\mathbf{I} - n^{-1}\,\mathbf{1}\mathbf{1}^{\mathrm{T}})$$

$$= \begin{pmatrix} 1 & 0 & 0 & 0 & 0 \\ 0 & 1 & 0 & 0 & 0 \\ 0 & 0 & 1 & 0 & 0 \\ 0 & 0 & 0 & 1 & 0 \\ 0 & 0 & 0 & 0 & 1 \end{pmatrix} - \frac{1}{5}\begin{pmatrix} 1 \\ 1 \\ 1 \\ 1 \\ 1 \end{pmatrix}(1 \quad 1 \quad 1 \quad 1 \quad 1)$$

$$= \begin{pmatrix} 1 & 0 & 0 & 0 & 0 \\ 0 & 1 & 0 & 0 & 0 \\ 0 & 0 & 1 & 0 & 0 \\ 0 & 0 & 0 & 1 & 0 \\ 0 & 0 & 0 & 0 & 1 \end{pmatrix} - \frac{1}{5}\begin{pmatrix} 1 & 1 & 1 & 1 & 1 \\ 1 & 1 & 1 & 1 & 1 \\ 1 & 1 & 1 & 1 & 1 \\ 1 & 1 & 1 & 1 & 1 \\ 1 & 1 & 1 & 1 & 1 \end{pmatrix}$$

$$= \begin{pmatrix} 0.8 & -0.2 & -0.2 & -0.2 & -0.2 \\ -0.2 & 0.8 & -0.2 & -0.2 & -0.2 \\ -0.2 & -0.2 & 0.8 & -0.2 & -0.2 \\ -0.2 & -0.2 & -0.2 & 0.8 & -0.2 \\ -0.2 & -0.2 & -0.2 & -0.2 & 0.8 \end{pmatrix},$$

from which we compute the centered matrix as

$$\tilde{\mathbf{X}} = \mathbf{C}\mathbf{X}$$

$$= \begin{pmatrix} 0.8 & -0.2 & -0.2 & -0.2 & -0.2 \\ -0.2 & 0.8 & -0.2 & -0.2 & -0.2 \\ -0.2 & -0.2 & 0.8 & -0.2 & -0.2 \\ -0.2 & -0.2 & -0.2 & 0.8 & -0.2 \\ -0.2 & -0.2 & -0.2 & -0.2 & 0.8 \end{pmatrix}\begin{pmatrix} 178 & 76 \\ 180 & 80 \\ 170 & 70 \\ 172 & 74 \\ 175 & 75 \end{pmatrix} = \begin{pmatrix} 3 & 1 \\ 5 & 5 \\ -5 & -5 \\ -3 & -1 \\ 0 & 0 \end{pmatrix}, \qquad (8.27)$$

which is the same as $\tilde{\mathbf{X}}$ in (8.23).

The centering matrix $\mathbf{C}$ has two important properties: $\mathbf{C}^{\mathrm{T}} = \mathbf{C}$ and $\mathbf{C}\mathbf{C} = \mathbf{C}$, that is, $\mathbf{C}$ is both symmetric and idempotent. In fact, both $n^{-1}\,\mathbf{1}\mathbf{1}^{\mathrm{T}}$ and $\mathbf{C}$ are projection matrices; the former projects onto the space generated by the vector $\mathbf{1}$ and the latter projects onto the orthogonal complement (see Section 7.5).

Now, having computed the centered matrix, the variance matrix can be written as

$$\begin{aligned} \mathbf{S} &= (n-1)^{-1}\,\tilde{\mathbf{X}}^{\mathrm{T}}\tilde{\mathbf{X}} \\ &= (n-1)^{-1}\,\mathbf{X}^{\mathrm{T}}\mathbf{C}^{\mathrm{T}}\mathbf{C}\mathbf{X} \qquad (8.28) \\ &= (n-1)^{-1}\,\mathbf{X}^{\mathrm{T}}\mathbf{C}\mathbf{X}. \end{aligned}$$

The last line follows because **C** is symmetric and idempotent. The matrix $\mathbf{X}^T\mathbf{X}$ is called the *sums of squares and cross products* matrix and the matrix $\tilde{\mathbf{X}}^T\tilde{\mathbf{X}}$ is called the *mean-adjusted sums of squares and cross products* matrix. Thus, the formula in (8.28) is an alternative way to compute the variance matrix. Note that the last formula for **S** in (8.28) involves neither $\bar{\mathbf{x}}$ nor $\tilde{\mathbf{X}}$. Thus, using **C** we can compute **S** directly from **X**. For example, using (8.28) we can compute the variance matrix of **X** in (8.14) as follows:

$$\mathbf{S} = (n-1)^{-1}\,\mathbf{X}^T\mathbf{C}\mathbf{X}$$

$$= \frac{1}{4}\begin{pmatrix} 178 & 180 & 170 & 172 & 175 \\ 76 & 80 & 70 & 74 & 75 \end{pmatrix}\begin{pmatrix} 0.8 & -0.2 & -0.2 & -0.2 & -0.2 \\ -0.2 & 0.8 & -0.2 & -0.2 & -0.2 \\ -0.2 & -0.2 & 0.8 & -0.2 & -0.2 \\ -0.2 & -0.2 & -0.2 & 0.8 & -0.2 \\ -0.2 & -0.2 & -0.2 & -0.2 & 0.8 \end{pmatrix}\begin{pmatrix} 178 & 76 \\ 180 & 80 \\ 170 & 70 \\ 172 & 74 \\ 175 & 75 \end{pmatrix}$$

$$= \begin{pmatrix} 17 & 14 \\ 14 & 13 \end{pmatrix},$$

which is the same as **S** in (8.18).

8.2.3. The Correlation Matrix

Another useful summary statistic is the *correlation coefficient*. The correlation coefficient is a measure of linear association between two variables. Given two variables $\mathbf{x}_j$ and $\mathbf{x}_l$, the correlation coefficient (also known as the *Pearson product-moment correlation coefficient*) between $\mathbf{x}_j$ and $\mathbf{x}_l$ is defined as

$$r_{jl} = \frac{s_{jl}}{s_{jj} \times s_{ll}}, \qquad (8.29)$$

which is the covariance between $\mathbf{x}_j$ and $\mathbf{x}_l$ divided by the product of the standard deviations of $\mathbf{x}_j$ and $\mathbf{x}_l$.

One property of the correlation coefficient is that it is invariant to the scale of measurements of the two variables. Thus, the correlation coefficient between $a\mathbf{x}_j$ and $b\mathbf{x}_l$ is the same as the correlation coefficient between $\mathbf{x}_j$ and $\mathbf{x}_l$, for any positive scalars a and b. For example, the correlation coefficient between salary and experience is the same whether we measure salary in dollars or thousands of dollars and/or we measure experience in months or in years.

The correlation coefficient also satisfies $-1 \leq r_{jl} \leq 1$. A positive value of r_{jl} indicates that the two variables are positively correlated, i.e., large (small) values of one variable are associated with large (small) values of the other. Negative values indicate the opposite relationship.

The correlation coefficients for a set of variables can be organized in a matrix,

$$\mathbf{R} = \begin{pmatrix} 1 & r_{12} & \cdots & r_{1k} \\ r_{21} & 1 & \cdots & r_{2k} \\ \vdots & \vdots & \ddots & \vdots \\ r_{k1} & r_{k2} & \cdots & 1 \end{pmatrix}, \tag{8.30}$$

which is known as the *correlation matrix*. The diagonal elements of the correlation matrix are 1, because a variable is perfectly correlated with itself. The matrix **R** can be expressed in terms of the variance-covariance matrix as

$$\mathbf{R} = \mathbf{DSD}, \tag{8.31}$$

where **S** is the variance-covariance matrix and **D** is a diagonal matrix with $d_{jj} = 1/s_{jj}$, the reciprocal of the standard deviation of $\mathbf{x}_j$.

As an example, the correlation matrix corresponding to **X** in (8.14) can be computed using (8.31) as follows:

$$\mathbf{R} = \begin{pmatrix} 1/\sqrt{17} & 0 \\ 0 & 1/\sqrt{13} \end{pmatrix} \begin{pmatrix} 17 & 14 \\ 14 & 13 \end{pmatrix} \begin{pmatrix} 1/\sqrt{17} & 0 \\ 0 & 1/\sqrt{13} \end{pmatrix} = \begin{pmatrix} 1.00 & 0.94 \\ 0.94 & 1.00 \end{pmatrix}.$$

Thus, the correlation coefficient between weights and heights is 0.94, which indicates that weight and height are positively and highly correlated, as would be expected.

*8.3. Regression Analysis

8.3.1. Introduction

Regression analysis provides a set of statistical tools for studying the relationship between two sets of variables; one set is regarded as dependent on or determined by the other set. For example, suppose that we would like to know what factors determine the price of a house. A real estate expert would tell us that the price of a house is determined by many factors, such as the lot size, the number and size of the bedrooms, the number and size of the bathrooms, the finished area, the garage size, the age and style, the location, and so on. We refer to the variable of interest (here the price of a house) as the *dependent* variable and to the variables thought to determine the dependent variable as the *predictors*, because they are usually used to predict the price of the house. The dependent variable is denoted by y and the predictors are denoted by $x_1, x_2, \ldots, x_k$, where k is the number of predictors.

The relationship between y and $x_1, x_2, \ldots, x_k$ can take many forms. The most commonly used form is

$$y = \beta_1 x_1 + \beta_2 x_2 + \cdots + \beta_k x_k + \varepsilon. \tag{8.32}$$

The β_1, β_2, …, and β_k are called the *regression coefficients* and ε is a *random error*, which can also be thought of as representing all the factors other than x_1, x_2, …, x_k on which y may depend. Equation (8.32) is called a *linear regression model* because it is linear in the regression coefficients.

There are several variations of the regression model in (8.32). Perhaps the most common regression model is the one in which one of the x-variables (usually x_1) is a constant (usually set equal to 1). In this case, (8.32) is referred to as a model with a *constant term* or a model with an *intercept*. This is because when $x_1 = 1$, β_1 can be interpreted as an intercept, namely, the value of y when all other x-variables are set to 0. If none of the x-variables are constants, (8.32) is referred to as a *no-intercept* model or a model *through the origin*, because y is 0 whenever all the x-variables are set to 0.

When (8.32) contains only one nonconstant x-variable, it is referred to as a *simple regression* model; otherwise it is referred to as a *multiple regression* model. The results below are applicable to both simple and multiple regressions. They are also applicable whether or not the model contains a constant term. (See Exercises 8.12–8.18.)

The regression coefficients in (8.32) are unknown parameters and one objective of the regression analysis is to estimate them. To do this, we collect a sample of n houses, and for each house we measure y, x_1, x_2, …, and x_k. Thus, for the ith house, the regression model in (8.32) can be written as

$$y_i = \beta_1 x_{i1} + \beta_2 x_{i2} + \cdots + \beta_k x_{ik} + \varepsilon_i, \quad i = 1, 2, \ldots, n, \tag{8.33}$$

where, for example, y_i is the price of the ith house, x_{i1} is the lot size of the ith house, x_{i2} is the number of bedrooms in the ith house, and so on. Our data can be organized as follows:

$$\mathbf{y} = \begin{pmatrix} y_1 \\ y_2 \\ \vdots \\ y_n \end{pmatrix} \quad \text{and} \quad \mathbf{X} = \begin{pmatrix} x_{11} & x_{12} & \cdots & x_{1k} \\ x_{21} & x_{22} & \cdots & x_{2k} \\ \vdots & \vdots & \ddots & \vdots \\ x_{n1} & x_{n2} & \cdots & x_{nk} \end{pmatrix}. \tag{8.34}$$

The vector $\mathbf{y}$ represents the prices for the n houses. The matrix $\mathbf{X}$ consists of n rows and k columns. Each column of $\mathbf{X}$ represents the values of one predictor for all n houses. Each row of $\mathbf{X}$ represents the values of all predictors for one house. Putting β_1, β_2, …, β_k and ε_1, ε_2, …, ε_n in two column vectors,

$$\boldsymbol{\beta} = \begin{pmatrix} \beta_1 \\ \beta_2 \\ \vdots \\ \beta_k \end{pmatrix} \quad \text{and} \quad \boldsymbol{\varepsilon} = \begin{pmatrix} \varepsilon_1 \\ \varepsilon_2 \\ \vdots \\ \varepsilon_n \end{pmatrix}, \tag{8.35}$$

the reader can verify that the regression model in (8.33) can be written as

$$\mathbf{y} = \mathbf{X}\boldsymbol{\beta} + \boldsymbol{\varepsilon}. \tag{8.36}$$

Thus, any linear regression model can always be written in the compact and simple form of (8.36), regardless of the number of predictors or the sample size.

8.3.2. Least Squares Estimates

Now we wish to estimate the regression coefficient vector $\boldsymbol{\beta}$ based on the data $\mathbf{y}$ and $\mathbf{X}$. The most popular method for estimating $\boldsymbol{\beta}$ is known as the *least squares* method. It is developed as follows. Equation (8.36) can also be written as $\boldsymbol{\varepsilon} = \mathbf{y} - \mathbf{X}\boldsymbol{\beta}$. According to the least squares method, we find the vector $\boldsymbol{\beta}$ that minimizes the sum of the squared elements of $\boldsymbol{\varepsilon}$, namely,

$$\sum_{i=1}^{n} \varepsilon_i^2 = \boldsymbol{\varepsilon}^T\boldsymbol{\varepsilon} = (\mathbf{y} - \mathbf{X}\boldsymbol{\beta})^T(\mathbf{y} - \mathbf{X}\boldsymbol{\beta}). \tag{8.37}$$

One way to do this is to find the derivative of (8.37) with respect to $\boldsymbol{\beta}$, set the derivative equal to $\mathbf{0}$, and solve for $\boldsymbol{\beta}$.[2] This yields

$$\mathbf{X}^T\mathbf{X}\boldsymbol{\beta} = \mathbf{X}^T\mathbf{y}. \tag{8.38}$$

Notice that this equation has the same form $\mathbf{Ax} = \mathbf{b}$ that we saw earlier in (8.3); thus, (8.38) represents a system of simultaneous linear equations. By comparison with (8.3), we see that $\boldsymbol{\beta}$ plays the role of $\mathbf{x}$, $\mathbf{X}^T\mathbf{X}$ plays the role of $\mathbf{A}$, and $\mathbf{X}^T\mathbf{y}$ plays the role of $\mathbf{b}$. Therefore, (8.38) is a system of k linear equations in k unknowns. These equations are referred to as the *normal* equations.

We now need to solve (8.38) for $\boldsymbol{\beta}$. With reference to Result 8.1, the augmented matrix for this system is $(\mathbf{X}^T\mathbf{X} : \mathbf{X}^T\mathbf{y}) = \mathbf{X}^T(\mathbf{X} : \mathbf{y})$. By Result 3.5(e),

$$\mathrm{r}(\mathbf{X}^T\mathbf{X} : \mathbf{X}^T\mathbf{y}) = \mathrm{r}[\mathbf{X}^T(\mathbf{X} : \mathbf{y})] \le \min[\mathrm{r}(\mathbf{X}^T), \mathrm{r}(\mathbf{X} : \mathbf{y})] = \mathrm{r}(\mathbf{X}),$$

because $\mathrm{r}(\mathbf{X}^T) = \mathrm{r}(\mathbf{X})$ and $\mathrm{r}(\mathbf{X}) \le \mathrm{r}(\mathbf{X} : \mathbf{y})$. Also by Result 3.5(e),

$$\mathrm{r}(\mathbf{X}^T\mathbf{X}) \le \min[\mathrm{r}(\mathbf{X}^T), \mathrm{r}(\mathbf{X})] = \min[\mathrm{r}(\mathbf{X}), \mathrm{r}(\mathbf{X})] = \mathrm{r}(\mathbf{X}).$$

Therefore, $\mathrm{r}(\mathbf{X}^T\mathbf{X} : \mathbf{X}^T\mathbf{y}) = \mathrm{r}(\mathbf{X}^T\mathbf{X})$; hence by Result 8.1, the system in (8.38) is consistent, that is, it has at least one solution. Furthermore, by Result 8.1(b), the system will have a unique solution if $\mathrm{r}(\mathbf{X}^T\mathbf{X}) = k$ (recall that k is the number of columns in $\mathbf{X}$); otherwise, the system has infinitely many solutions. For this reason it is common in applications of the least squares method to require that $\mathrm{r}(\mathbf{X}) = k$, namely, that $\mathbf{X}$ be of full-column rank or, equivalently, that $\mathbf{X}$ consists of k linearly independent columns.

It follows from (8.9) that, if $\mathrm{r}(\mathbf{X}) = k$, the system in (8.38) has a unique solution, which is given by

$$\hat{\boldsymbol{\beta}} = (\mathbf{X}^T\mathbf{X})^{-1}\mathbf{X}^T\mathbf{y}. \tag{8.39}$$

---

2 If you do not know how to find the derivative, you can simply take (8.38) on faith.

Notice that we replace β by $\hat{\beta}$ (read as beta-hat) to indicate that $\hat{\beta}$ is an estimate of β based on the data at hand. The vector $\hat{\beta}$ is the least squares estimate of β.

If **X** is rank deficient, the system (8.38) will have infinitely many solutions. It follows from (8.11) that one particular solution can be obtained as

$$\hat{\beta} = (\mathbf{X}^T\mathbf{X})^{-}\mathbf{X}^T\mathbf{y}, \tag{8.40}$$

where $(\mathbf{X}^T\mathbf{X})^{-}$ is any g-inverse of $(\mathbf{X}^T\mathbf{X})$.

8.3.3. An Illustrative Example

Suppose we have a sample of only five houses, a small enough sample for the computations to be carried out by hand. Further, suppose we consider only the following variables:

y = the price in thousands of dollars,
x_1 = lot size (in hundreds of square yards),
x_2 = number of bedrooms.

The linear model in (8.32) becomes

$$y = \beta_1 x_1 + \beta_2 x_2 + \varepsilon, \tag{8.41}$$

and the data can be organized as follows:

$$\mathbf{y} = \begin{pmatrix} 122 \\ 115 \\ 128 \\ 145 \\ 108 \end{pmatrix} \quad \text{and} \quad \mathbf{X} = \begin{pmatrix} 10 & 4 \\ 12 & 3 \\ 12 & 4 \\ 14 & 5 \\ 10 & 3 \end{pmatrix}. \tag{8.42}$$

In this example, **y** is 5×1, **X** is 5×2, and

$$\beta = \begin{pmatrix} \beta_1 \\ \beta_2 \end{pmatrix}.$$

Thus, model (8.41) can be written in the form given in (8.36).

To compute the least squares estimate of β, we need first to compute $(\mathbf{X}^T\mathbf{X})$, $(\mathbf{X}^T\mathbf{X})^{-1}$, and $\mathbf{X}^T\mathbf{y}$. Note that the matrices **y** and **X** in (8.42) are the same as those in (7.19). So, we have already computed $(\mathbf{X}^T\mathbf{X})$, $(\mathbf{X}^T\mathbf{X})^{-1}$ in (7.20) and (7.21). We then compute

$$\mathbf{X}^{\mathrm{T}}\mathbf{y} = \begin{pmatrix} 10 & 12 & 12 & 14 & 10 \\ 4 & 3 & 4 & 5 & 3 \end{pmatrix} \begin{pmatrix} 122 \\ 115 \\ 128 \\ 145 \\ 108 \end{pmatrix} = \begin{pmatrix} 7246 \\ 2394 \end{pmatrix}.$$

Substituting these matrices into (8.39), we obtain

$$\hat{\beta} = (\mathbf{X}^{\mathrm{T}}\mathbf{X})^{-1}\mathbf{X}^{\mathrm{T}}\mathbf{y}$$

$$= \frac{1}{1124}\begin{pmatrix} 75 & -224 \\ -224 & 684 \end{pmatrix} \begin{pmatrix} 7246 \\ 2394 \end{pmatrix} = \begin{pmatrix} 6.400 \\ 12.804 \end{pmatrix}. \tag{8.43}$$

Thus, $\hat{\beta}_1 = 6.4$ and $\hat{\beta}_2 = 12.8$. An interpretation of these values is given next.

8.3.4. The Fitted Values and Residual Vectors

Once an estimate of the regression coefficients has been obtained, one can then estimate the price of a house given its number of bedrooms and lot size. Thus, to estimate the price of a house for which lot size and number of bedrooms are represented by x_1 and x_2, respectively, we use

$$\hat{y} = 6.4\, x_1 + 12.8\, x_2.$$

Recall that the price is measured in thousands of dollars and the lot size is measured in hundreds of square yards. Thus, thě above equation implies that, holding other things constant, each extra 100 square yards in the lot size will increase the estimated price of the house by \$6,400 (\$1000 for each unit of x_1) or, equivalently, each extra square yard in the lot size will increase the estimated price of the house by \$64. Similarly, holding other things constant, each extra bedroom will increase the estimated price by \$12,800. Thus, for example, an estimate of the price of the second house in the sample (a house with three bedrooms and a lot size of 1200 square yards) is \$115,200 because

$$\hat{y}_2 = 6.4 \times 12 + 12.8 \times 3 = 115.2,$$

whereas an estimate of the price of the third house (with four bedrooms and the same lot size) is \$128,000 because

$$\hat{y}_3 = 6.4 \times 12 + 12.8 \times 4 = 128.$$

The difference of \$12,800 between the estimated price of the above two houses is due to the fact that the third house has one more bedroom than the second.

The vector $\hat{\mathbf{y}} = \mathbf{X}\hat{\boldsymbol{\beta}}$ is called the vector of *fitted* or *predicted values*. The difference between the actual and fitted values , namely,

$$\mathbf{e} = \mathbf{y} - \hat{\mathbf{y}}, \tag{8.44}$$

is called the *residuals* or *vector of residuals*. For our example, the fitted values and the residuals are

$$\hat{\mathbf{y}} = \mathbf{X}\hat{\boldsymbol{\beta}} = \begin{pmatrix} 10 & 4 \\ 12 & 3 \\ 12 & 4 \\ 14 & 5 \\ 10 & 3 \end{pmatrix} \begin{pmatrix} 6.4 \\ 12.8 \end{pmatrix} = \begin{pmatrix} 115.2 \\ 115.2 \\ 128.0 \\ 153.6 \\ 102.4 \end{pmatrix}, \tag{8.45}$$

and

$$\mathbf{e} = \mathbf{y} - \hat{\mathbf{y}} = \begin{pmatrix} 122 \\ 115 \\ 128 \\ 145 \\ 108 \end{pmatrix} - \begin{pmatrix} 115.2 \\ 115.2 \\ 128.0 \\ 153.6 \\ 102.4 \end{pmatrix} = \begin{pmatrix} 6.8 \\ -0.2 \\ 0.0 \\ -8.6 \\ 5.6 \end{pmatrix}. \tag{8.46}$$

From the vector of residuals we see, for example, that the price of the first house is underestimated by \$6,800; that the price of the second house is overestimated by \$200; that the estimated price of the third house is just right; and so on.

When $\mathbf{X}$ is of full-column rank, the fitted values can alternatively be written as

$$\begin{aligned} \hat{\mathbf{y}} &= \mathbf{X}\hat{\boldsymbol{\beta}} \\ &= \mathbf{X}(\mathbf{X}^T\mathbf{X})^{-1}\mathbf{X}^T\mathbf{y} \\ &= \mathbf{P}\mathbf{y}, \end{aligned} \tag{8.47}$$

where

$$\mathbf{P} = \mathbf{X}(\mathbf{X}^T\mathbf{X})^{-1}\mathbf{X}^T. \tag{8.48}$$

The vector of residuals can also be written as

$$\mathbf{e} = \mathbf{y} - \hat{\mathbf{y}} = \mathbf{y} - \mathbf{P}\mathbf{y} = (\mathbf{I}_n - \mathbf{P})\mathbf{y}. \tag{8.49}$$

As can be seen from (8.47) and (8.49), the $n \times n$ matrix $\mathbf{P}$ determines both the vector of fitted values and the vector of residuals. Both $\mathbf{P}$ and $(\mathbf{I}_n - \mathbf{P})$ in (8.48) and (8.49) are projection matrices.[3] Therefore, the vector of fitted values $\hat{\mathbf{y}}$ can be interpreted as the orthogonal projection of $\mathbf{y}$ onto the column space of $\mathbf{X}$. Similarly, the vector of residuals $\mathbf{e}$ can be interpreted as the orthogonal projection of $\mathbf{y}$ onto the

3 See Section 7.5 for the definition of projection matrices.

orthogonal complement of **X**. Also, by Result 7.5, we have that **P** and $(\mathbf{I}_n - \mathbf{P})$ are symmetric and idempotent matrices.[4] These properties, for example, allow us to express the sum of squared residuals as

$$\begin{aligned}\mathbf{e}^T\mathbf{e} &= \mathbf{y}^T(\mathbf{I}_n - \mathbf{P})^T(\mathbf{I}_n - \mathbf{P})\mathbf{y} \\ &= \mathbf{y}^T(\mathbf{I}_n - \mathbf{P})(\mathbf{I}_n - \mathbf{P})\mathbf{y} \qquad \text{(because of symmetry)} \\ &= \mathbf{y}^T(\mathbf{I}_n - \mathbf{P})\mathbf{y} \qquad \text{(because of idempotency).} \qquad (8.50)\end{aligned}$$

Additionally, by Result 7.5, we have $\text{tr}(\mathbf{P}) = \text{r}(\mathbf{X}) = k$, and $\text{tr}(\mathbf{I}_n - \mathbf{P}) = n - k$. The matrices **P** and $(\mathbf{I}_n - \mathbf{P})$ possess many other interesting properties. For a comprehensive account of these properties, see Hadi (1986) or Chatterjee and Hadi (1988). As can be seen from the above, the projection matrices play an important role in linear regression.

We should emphasize here that this section serves only as an example illustrating how matrix algebra can be used to simplify otherwise very complicated results. Regression analysis is a vast topic; indeed, many volumes have been written on the subject. For more details, the interested reader is referred to books such as Seber (1977); Weisberg (1980); Draper and Smith (1981); Fox (1984); Neter, Wasserman, and Kutner (1985); Sen and Srivastava (1990); and Chatterjee and Price (1991), to mention only a few.

*8.4. Updating Computations

In this section we give two applications of the updating formulas of Section 6.2.4. The first example illustrates the case where we add variables (columns) to the data matrix **X**, and the second illustrates the case where we add observations (rows) to **X**.

8.4.1. Adding Columns to a Data Matrix

Suppose that after the regression calculations in the previous section have been completed, some new information has become available. We wish to take the new information into account. Specifically, suppose that data for a new variable, x_3 = number of bathrooms, are now available for the five houses. The new data are organized in the vector

$$\mathbf{x}_3 = \begin{pmatrix} 1 \\ 2 \\ 2 \\ 2 \\ 1 \end{pmatrix}.$$

The new model can be expressed as

$$y = \beta_1 x_1 + \beta_2 x_2 + \beta_3 x_3 + \varepsilon.$$

4 Recall from Section 2.3.2 that a matrix **P** satisfying **PP** = **P** is called an idempotent matrix.

Augmenting $\mathbf{x}_3$ to $\mathbf{X}$ in (8.42) we obtain

$$\mathbf{Z} = (\mathbf{X} : \mathbf{x}_3) = \begin{pmatrix} 10 & 4 & \vdots & 1 \\ 12 & 3 & \vdots & 2 \\ 12 & 4 & \vdots & 2 \\ 14 & 5 & \vdots & 2 \\ 10 & 3 & \vdots & 1 \end{pmatrix}.$$

We can, of course, start the regression calculations from scratch. Alternatively, we may make use of some of the calculations that we have already done. For example, we have

$$\mathbf{Z}^{\mathrm{T}}\mathbf{Z} = \begin{pmatrix} \mathbf{X}^{\mathrm{T}}\mathbf{X} & \mathbf{X}^{\mathrm{T}}\mathbf{x}_3 \\ \mathbf{x}_3^{\mathrm{T}}\mathbf{X} & \mathbf{x}_3^{\mathrm{T}}\mathbf{x}_3 \end{pmatrix} = \begin{pmatrix} 684 & 224 & 96 \\ 224 & 75 & 31 \\ 96 & 31 & 14 \end{pmatrix}. \tag{8.51}$$

This is a partitioned matrix and its inverse can be found easily by using Result 6.4. The partitioned matrix in (8.51) is already in the form given by (6.8), where

$$\mathbf{A} = \mathbf{X}^{\mathrm{T}}\mathbf{X} = \begin{pmatrix} 684 & 224 \\ 224 & 75 \end{pmatrix}, \quad \mathbf{b} = \mathbf{c} = \mathbf{X}^{\mathrm{T}}\mathbf{x}_3 = \begin{pmatrix} 96 \\ 31 \end{pmatrix}, \quad d = \mathbf{x}_3^{\mathrm{T}}\mathbf{x}_3 = 14.$$

Hence $(\mathbf{Z}^{\mathrm{T}}\mathbf{Z})^{-1}$ can be computed using either (6.9) or (6.10). Using (6.9), for example, we obtain

$$\mathbf{c}^{\mathrm{T}}\mathbf{A}^{-1}\mathbf{b} = (96 \quad 31)\begin{pmatrix} 684 & 224 \\ 224 & 75 \end{pmatrix}^{-1}\begin{pmatrix} 96 \\ 31 \end{pmatrix} = 13.5907,$$

$$h = (d - \mathbf{c}^{\mathrm{T}}\mathbf{A}^{-1}\mathbf{b})^{-1} = (14 - 13.5907)^{-1} = 2.4435,$$

$$\mathbf{g}^{\mathrm{T}} = -h\mathbf{c}^{\mathrm{T}}\mathbf{A}^{-1} = -2.4435(96 \quad 31)\begin{pmatrix} 684 & 224 \\ 224 & 75 \end{pmatrix}^{-1} = (-0.5565 \quad 0.6522),$$

$$\mathbf{f} = -\mathbf{A}^{-1}\mathbf{b}h = -2.4435\begin{pmatrix} 684 & 224 \\ 224 & 75 \end{pmatrix}^{-1}\begin{pmatrix} 96 \\ 31 \end{pmatrix} = \begin{pmatrix} -0.5565 \\ 0.6522 \end{pmatrix},$$

$$\mathbf{E} = \mathbf{A}^{-1} - \mathbf{f}\mathbf{c}^{\mathrm{T}}\mathbf{A}^{-1}$$

$$= \begin{pmatrix} 684 & 224 \\ 224 & 75 \end{pmatrix}^{-1} - \begin{pmatrix} -0.5565 \\ 0.6522 \end{pmatrix}(96 \quad 31)\begin{pmatrix} 684 & 224 \\ 224 & 75 \end{pmatrix}^{-1}$$

$$= \begin{pmatrix} 0.1935 & -0.3478 \\ -0.3478 & 0.7826 \end{pmatrix}.$$

Therefore,

$$(\mathbf{Z}^T\mathbf{Z})^{-1} = \begin{pmatrix} \mathbf{E} & \mathbf{f} \\ \mathbf{g}^T & h \end{pmatrix} = \begin{pmatrix} 0.1935 & -0.3478 & -0.5565 \\ -0.3478 & 0.7826 & 0.6522 \\ -0.5565 & 0.6522 & 2.4435 \end{pmatrix}, \qquad (8.52)$$

to the nearest four decimal places. You may verify that $(\mathbf{Z}^T\mathbf{Z})^{-1}$ is indeed the inverse of $\mathbf{Z}^T\mathbf{Z}$ by multiplying the two matrices to obtain $\mathbf{I}_3$. You now know how to find the inverse of a 3×3 matrix using only the formula for the inverse of a 2×2 matrix which is given in (6.4).

Now that we have computed $(\mathbf{Z}^T\mathbf{Z})^{-1}$, we can continue with the rest of the regression calculations as in the previous section. In particular, you may use (8.39) to verify that the new estimated regression coefficients are

$$\hat{\beta} = (\mathbf{Z}^T\mathbf{Z})^{-1}\mathbf{Z}^T\mathbf{y} = \begin{pmatrix} 9.387 \\ 9.304 \\ -13.113 \end{pmatrix}.$$

8.4.2. Adding Rows to a Data Matrix

Suppose now that the new information that has become available is about an additional house (observation or row) that has five bedrooms, a lot size of 1500 square yards, and a price of \$150,000. Thus, for this house $y = 150$, $x_1 = 15$, and $x_2 = 5$. Let us put x_1 and x_2 in a row vector $\mathbf{v}^T = (15 \quad 5)$, and use this row vector to augment the matrix $\mathbf{X}$ in (8.42). The new matrix becomes

$$\mathbf{W} = \begin{pmatrix} \mathbf{X} \\ \mathbf{v}^T \end{pmatrix} = \begin{pmatrix} 10 & 4 \\ 12 & 3 \\ 12 & 4 \\ 14 & 5 \\ 10 & 3 \\ 15 & 5 \end{pmatrix}.$$

Hence,

$$\mathbf{W}^T\mathbf{W} = (\mathbf{X}^T : \mathbf{v})\begin{pmatrix} \mathbf{X} \\ \mathbf{v}^T \end{pmatrix} = \mathbf{X}^T\mathbf{X} + \mathbf{v}\mathbf{v}^T$$

$$= \begin{pmatrix} 684 & 224 \\ 224 & 75 \end{pmatrix} + \begin{pmatrix} 15 \\ 5 \end{pmatrix}(15 \quad 5) \qquad (8.53)$$

$$= \begin{pmatrix} 909 & 299 \\ 299 & 100 \end{pmatrix}. \tag{8.54}$$

The matrices in (8.53) are in a form ready for the application of (6.12), where

$$\mathbf{A} = \mathbf{X}^{\mathrm{T}}\mathbf{X} = \begin{pmatrix} 684 & 224 \\ 224 & 75 \end{pmatrix} \quad \text{and} \quad \mathbf{b} = \mathbf{c} = \mathbf{v} = \begin{pmatrix} 15 \\ 5 \end{pmatrix}.$$

Thus,

$$\mathbf{A}^{-1}\mathbf{b} = \begin{pmatrix} 684 & 224 \\ 224 & 75 \end{pmatrix}^{-1} \begin{pmatrix} 15 \\ 5 \end{pmatrix} = \begin{pmatrix} 0.0044 \\ 0.0534 \end{pmatrix},$$

$$\mathbf{c}^{\mathrm{T}}\mathbf{A}^{-1} = (\mathbf{A}^{-1}\mathbf{b})^{\mathrm{T}} = \begin{pmatrix} 0.0044 & 0.0534 \end{pmatrix},$$

$$\mathbf{c}^{\mathrm{T}}\mathbf{A}^{-1}\mathbf{b} = \begin{pmatrix} 15 & 5 \end{pmatrix} \begin{pmatrix} 684 & 224 \\ 224 & 75 \end{pmatrix}^{-1} \begin{pmatrix} 15 \\ 5 \end{pmatrix} = 0.3336.$$

Therefore,

$$(\mathbf{W}^{\mathrm{T}}\mathbf{W})^{-1} = (\mathbf{A} + \mathbf{b}\mathbf{c}^{\mathrm{T}})^{-1} = \mathbf{A}^{-1} - \frac{\mathbf{A}^{-1}\mathbf{b}\mathbf{c}^{\mathrm{T}}\mathbf{A}^{-1}}{1 + \mathbf{c}^{\mathrm{T}}\mathbf{A}^{-1}\mathbf{b}}$$

$$= \begin{pmatrix} 684 & 224 \\ 224 & 75 \end{pmatrix}^{-1} - \frac{1}{1.3336} \begin{pmatrix} 0.0044 \\ 0.0534 \end{pmatrix} \begin{pmatrix} 0.0044 & 0.0534 \end{pmatrix}$$

$$= \begin{pmatrix} 0.0667 & -0.1994 \\ -0.1994 & 0.6064 \end{pmatrix}.$$

Of course, in this case it would be easier to compute $(\mathbf{W}^{\mathrm{T}}\mathbf{W})^{-1}$ directly from $\mathbf{W}^{\mathrm{T}}\mathbf{W}$ in (8.54) using formula (6.4). However, substantial computational savings can be gained when the matrix $\mathbf{X}^{\mathrm{T}}\mathbf{X}$ is large, because we have already computed $(\mathbf{W}^{\mathrm{T}}\mathbf{W})^{-1}$ and no further matrix inversions are necessary.

The rest of the regression calculations can be carried out as before. The new estimate of the regression coefficients is found to be

$$\hat{\beta} = (\mathbf{W}^{\mathrm{T}}\mathbf{W})^{-1}\mathbf{W}^{\mathrm{T}}\mathbf{y} = \begin{pmatrix} 6.367 \\ 12.403 \end{pmatrix}. \tag{8.55}$$

Compare the estimated coefficients in (8.43) and (8.55); the difference between the two is the effect of adding the sixth house to the sample. The rest of the regression calculations can be carried out as in the previous section.

We should point out here that the above updating formulas are useful not only for computational purposes, but also for deriving many theoretical results (see, for example, Chatterjee and Hadi, 1988).

Appendix: Solving Systems of Linear Equations

There are several methods for solving systems of linear equations. In this appendix we discuss one simple but general method. This method is based on the elementary row operations presented in the appendix to Chapter 3 and is very similar to the method for finding the inverse of a square matrix discussed in the appendix to Chapter 6.

Suppose for the moment that the system has a unique solution. To solve the system one may use the following steps:

Step 1. Create the augmented matrix $\mathbf{B} = (\mathbf{A} : \mathbf{b})$ as in (8.8). If $\mathbf{A}$ is an $m \times n$ matrix, then $\mathbf{B}$ is of order $m \times (n + 1)$, and each row of $\mathbf{B}$ represents one equation.

Step 2. Systematically apply a sequence of row operations to the augmented matrix to reduce the matrix $\mathbf{A}$ to $\mathbf{I}$. The same sequence of row operations will then replace $\mathbf{b}$ in the augmented matrix by a vector representing the solution to the system of linear equations.

Let us now try this algorithm on three different examples. The first is the system in (8.4) for which

$$\mathbf{A} = \begin{pmatrix} 1 & 1 \\ 1 & -1 \end{pmatrix}, \quad \mathbf{x} = \begin{pmatrix} x_1 \\ x_2 \end{pmatrix}, \quad \mathbf{b} = \begin{pmatrix} 5 \\ 1 \end{pmatrix}.$$

According to Step 1, the augmented matrix $\mathbf{B}$ is

$$\mathbf{B} = \begin{pmatrix} 1 & 1 & : & 5 \\ 1 & -1 & : & 1 \end{pmatrix}.$$

In Step 2, we perform the row operations necessary to reduce $\mathbf{A}$ to $\mathbf{I}$, as follows:

$$\mathbf{B}_1 = \begin{pmatrix} 1 & 1 & : & 5 \\ 0 & -2 & : & -4 \end{pmatrix} \begin{matrix} \mathbf{r}_1 \\ \mathbf{r}_2 - \mathbf{r}_1 \end{matrix},$$

$$\mathbf{B}_2 = \begin{pmatrix} 1 & 0 & : & 3 \\ 0 & 1 & : & 2 \end{pmatrix} \begin{matrix} \mathbf{r}_1 + 0.5\mathbf{r}_2 \\ -0.5\mathbf{r}_2 \end{matrix}.$$

Now, the matrix $\mathbf{A}$ has been successfully replaced by $\mathbf{I}$; hence the solution to (8.1) is the last vector in $\mathbf{B}_2$, namely,

$$\mathbf{x} = \begin{pmatrix} x_1 \\ x_2 \end{pmatrix} = \begin{pmatrix} 3 \\ 2 \end{pmatrix},$$

which is the same as the solution we obtained graphically in Figure 8.1 and the same as that in (8.10) using $\mathbf{A}^{-1}$.

Now, let us try to solve (8.5) using the above algorithm. The augmented matrix for this system is

$$\mathbf{B} = \begin{pmatrix} 1 & 1 & : & 5 \\ 1 & -1 & : & 1 \\ 2 & 2 & : & 10 \end{pmatrix}.$$

Performing the appropriate row operations, we obtain

$$\mathbf{B}_1 = \begin{pmatrix} 1 & 1 & : & 5 \\ 0 & -2 & : & -4 \\ 0 & 0 & : & 0 \end{pmatrix} \begin{matrix} \mathbf{r}_1 \\ \mathbf{r}_2 - \mathbf{r}_1 \\ \mathbf{r}_3 - 2\mathbf{r}_1 \end{matrix} .$$

The last row of $\mathbf{B}_1$ is a null vector, which indicates that the last equation in (8.5) is redundant. We therefore delete this row and continue the row operations. We obtain

$$\mathbf{B}_2 = \begin{pmatrix} 1 & 0 & : & 3 \\ 0 & 1 & : & 2 \end{pmatrix} \begin{matrix} \mathbf{r}_1 + 0.5\mathbf{r}_1 \\ -0.5\mathbf{r}_2 \end{matrix} ,$$

from which we see that the solution to (8.5) is $x_1 = 3$ and $x_2 = 2$.

Let us now see what happens when we apply the above algorithm to the linear system in (8.6). The corresponding augmented matrix is given by

$$\mathbf{B} = \begin{pmatrix} 1 & 1 & : & 5 \\ 1 & -1 & : & 1 \\ 1 & -4 & : & 0 \end{pmatrix}.$$

Performing row operations to reduce the first column of $\mathbf{B}$, we obtain

$$\mathbf{B}_1 = \begin{pmatrix} 1 & 1 & : & 5 \\ 0 & -2 & : & -4 \\ 0 & -5 & : & -5 \end{pmatrix} \begin{matrix} \mathbf{r}_1 \\ \mathbf{r}_2 - \mathbf{r}_1 \\ \mathbf{r}_3 - \mathbf{r}_1 \end{matrix} .$$

Performing row operations to reduce the second column of $\mathbf{B}_1$, we obtain

$$\mathbf{B}_2 = \begin{pmatrix} 1 & 0 & : & 3 \\ 0 & 1 & : & 2 \\ 0 & 0 & : & 5 \end{pmatrix} \begin{matrix} \mathbf{r}_1 + 0.5\mathbf{r}_2 \\ -0.5\mathbf{r}_2 \\ \mathbf{r}_3 - 2.5\mathbf{r}_2 \end{matrix} .$$

Notice that $\mathbf{B}_2$ is in row-echelon form and that $r(\mathbf{B}) = r(\mathbf{B}_2) = 3$. Since $r(\mathbf{A}) = 2$, this means that the vector $\mathbf{b}$ is independent of the two column vectors in $\mathbf{A}$, which indicates in turn that the system (8.6) is inconsistent.

Exercises

8.1. For each of the following systems, find a matrix $\mathbf{A}$ and a vector $\mathbf{b}$ so that the system can be written as $\mathbf{Ax} = \mathbf{b}$.
(a) $2x + y + z = 2$, $x + y + z = 0$, and $3x - 2y = 0$.
(b) $x + 2y + 3z = 3$, $2x + y + z = 0$, and $x + 4y + z = 1$.
(c) $3x - 2y = 1$, $y + z = 0$, and $x + z = 1$.

8.2. Using graphical and then numerical methods, find a solution, if any exists, for each of the following systems of linear equations and check that the two methods give the same answer.
(a) $x + y = 2$ and $x - y = -1$.
(b) $x + y = 2$ and $2x - 2y = 4$.
(c) $2x + 9y = 1$ and $3x - 2y = 4$.
(d) $x + 5y = 2$ and $2x + 10y = 3$.
(e) $3x + y = 7$ and $-6x - 2y = -14$.
(f) $x + y = 0$ and $x - y = 0$.
(g) $x + y = 0$ and $2x + 2y = 0$.

8.3. Two systems $\mathbf{Ax}_1 = \mathbf{b}_1$ and $\mathbf{Ax}_2 = \mathbf{b}_2$ can be written in one equation as $\mathbf{AX} = \mathbf{B}$, where $\mathbf{X} = (\mathbf{x}_1 : \mathbf{x}_2)$ and $\mathbf{B} = (\mathbf{b}_1 : \mathbf{b}_2)$. This can be generalized to more than two systems. Furthermore, all systems, if consistent, can be solved simultaneously by computing the regular or g-inverse of $\mathbf{A}$, then post-multiplying it by $\mathbf{B}$ to obtain $\mathbf{X} = \mathbf{A}^{-1}\mathbf{B}$ or $\mathbf{X} = \mathbf{A}^{-}\mathbf{B}$.
(a) What are the conditions under which the system $\mathbf{AX} = \mathbf{B}$ is consistent?
(b) Specify the conditions under which $\mathbf{AX} = \mathbf{B}$ has a unique solution.
(c) What are the conditions under which the system $\mathbf{AX} = \mathbf{B}$ is inconsistent while the system $\mathbf{Ax}_1 = \mathbf{b}_1$ is consistent?
(d) Solve the two systems specified by

$$\mathbf{A} = \begin{pmatrix} 1 & -1 \\ 1 & 1 \end{pmatrix}, \quad \mathbf{b}_1 = \begin{pmatrix} 2 \\ 1 \end{pmatrix}, \quad \text{and} \quad \mathbf{b}_1 = \begin{pmatrix} 1 \\ 2 \end{pmatrix}.$$

(e) Suppose that $\mathbf{x}_1 = \mathbf{x}_2$; what are the conditions under which the system $\mathbf{AX} = \mathbf{B}$ is consistent?

8.4. A person is considering each of two types of investment. Each week the first type pays \$480 plus 5 percent of the amount invested, while the second type just pays 9 percent per week. What is the minimum amount of money that must be invested in order that the two types of investment yield the same amount each week? (Adapted from Joseph and Styan, 1987.)

8.5. Let $\mathbf{A} = \begin{pmatrix} 2 & 0 & 4 \\ 0 & 1 & 0 \\ 1 & 1 & 0 \end{pmatrix}$ and $\mathbf{b} = \begin{pmatrix} 2 \\ 1 \\ 6 \end{pmatrix}$.

(a) Write the three equations represented by the system $\mathbf{Ax} = \mathbf{b}$.
(b) Find a vector $\mathbf{x}$ such that $\mathbf{Ax} = \mathbf{b}$.

8.6. Let $\mathbf{A} = \begin{pmatrix} 1 & s \\ 1 & 2 \end{pmatrix}$, $\mathbf{b} = \begin{pmatrix} 0 \\ 0 \end{pmatrix}$, $\mathbf{c} = \begin{pmatrix} 1 \\ 1 \end{pmatrix}$; find all values of s, if any, for which the system:
(a) $\mathbf{Ax} = \mathbf{b}$ has no solution.
(b) $\mathbf{Ax} = \mathbf{b}$ has a unique solution.
(c) $\mathbf{Ax} = \mathbf{c}$ has no solution.
(d) $\mathbf{Ax} = \mathbf{c}$ has a unique solution.
(e) $\mathbf{Ax} = \mathbf{c}$ has infinitely many solutions.

8.7. Each of the linear systems below may have either no solution or infinitely many solutions; determine which is the case for each system. If the system has infinitely many solutions, determine how many free variables there are and then find two different solutions.
(a) $x + y = 1$.
(b) $x + y = 1$, $-x - y = -1$, and $2x + 2y = 1$.
(c) $x + y + z = 5$, and $x - y - z = -4$.
(d) $x + y - z = 1$, $x - y + z = 4$, and $2x + y - z = 0$.

8.8. Consider the linear system $x + y - z = 1$ and $x - y + z = 4$. Show that, although the system has infinitely many solutions, one of the variables has a unique solution (which one?).

8.9. Let $\mathbf{1}$ be a column vector of n ones and $\mathbf{P}$ be the projection matrix for the space generated by $\mathbf{1}$.
(a) Show that $\mathbf{P} = n^{-1}\,\mathbf{1}\mathbf{1}^{\mathrm{T}}$, hence the centering matrix in (8.26) can be written as $\mathbf{C} = (\mathbf{I} - \mathbf{P})$.
(b) Show that $\mathbf{C1} = \mathbf{0}$.
(c) Use $\mathbf{P}$ and $\mathbf{C}$ above to verify Result 7.5(a)–(e).

8.10. For each of the following data matrices,

$$\mathbf{X} = \begin{pmatrix} 1 & -1 & 0 \\ 1 & 1 & 0 \\ -1 & 0 & 1 \\ -1 & 0 & 1 \end{pmatrix}, \quad \mathbf{Y} = \begin{pmatrix} 1 & -1 & 1 \\ 1 & 1 & 1 \\ -1 & 0 & 1 \\ -1 & 0 & 1 \end{pmatrix}, \quad \mathbf{Z} = \begin{pmatrix} 1 & -1 & 0 \\ 1 & 1 & 2 \\ -1 & 0 & -1 \\ -1 & 0 & -1 \end{pmatrix},$$

compute:
(a) the mean vector.
(b) the centered matrix.
(c) the variance-covariance matrix.
(d) the correlation matrix.
(e) Either prove (8.24) or verify it using the above matrices.

8.11. For each of the matrices in Exercise 8.10:
(a) Center the matrix by subtracting from each element the mean of its column.
(b) Divide each element of the centered matrix by its standard deviation. The resultant matrix is called the *standardized data*.
(c) Compute the variance-covariance matrix of the standardized data matrix.
(d) Verify that the covariance matrix obtained in (c) is the same as the correlation matrix of the original data matrix obtained in Exercise 8.10(d). (This leads to an interesting result which states that the correlation matrix is nothing more than the covariance matrix of the standardized data.)

8.12. Compute the least squares estimate of β obtained from fitting the linear model $\mathbf{y} = \mathbf{X}\beta + \varepsilon$ to the following data:

$$\text{(a) } \mathbf{X} = \begin{pmatrix} 1 & -1 \\ 1 & 1 \\ 1 & 0 \\ 1 & 0 \end{pmatrix} \quad \text{and} \quad \mathbf{y} = \begin{pmatrix} 2 \\ 2 \\ 0 \\ 0 \end{pmatrix}.$$

$$\text{(b) } \mathbf{X} = \begin{pmatrix} 1 & -1 & 1 \\ 1 & 1 & 1 \\ 1 & 0 & -1 \\ 1 & 0 & -1 \end{pmatrix} \quad \text{and} \quad \mathbf{y} = \begin{pmatrix} 2 \\ 2 \\ 0 \\ 0 \end{pmatrix}.$$

8.13. Consider the following simple regression model with an intercept:

$$y_i = \beta_0 + \beta_1 x_i + \varepsilon_i, \quad i = 1, 2, \ldots, n.$$

Here β_0 is the intercept and β_1 is the slope of the regression line. This model can be written in vector terms as $\mathbf{y} = \beta_0\mathbf{1} + \beta_1 \mathbf{x} + \varepsilon$, where $\mathbf{1}$ is a column vector of ones.
(a) Create a matrix $\mathbf{X}$ and a vector β so that the above model can also be written $\mathbf{y} = \mathbf{X}\beta + \varepsilon$, as in (8.36).
(b) Show that

$$\mathbf{X}^{\mathrm{T}}\mathbf{X} = \begin{pmatrix} n & \sum_{i=1}^{n} x_i \\ \sum_{i=1}^{n} x_i & \sum_{i=1}^{n} x_i^2 \end{pmatrix}, \quad (\mathbf{X}^{\mathrm{T}}\mathbf{X})^{-1} = \frac{1}{\det(\mathbf{X}^{\mathrm{T}}\mathbf{X})} \begin{pmatrix} \sum_{i=1}^{n} x_i^2 & -\sum_{i=1}^{n} x_i \\ -\sum_{i=1}^{n} x_i & n \end{pmatrix},$$

and

$$\mathbf{X}^{\mathrm{T}}\mathbf{y} = \begin{pmatrix} \sum_{i=1}^{n} y_i \\ \sum_{i=1}^{n} x_i y_i \end{pmatrix}.$$

(c) Compute the least squares estimate of β, $\hat{\beta} = \begin{pmatrix} \hat{\beta}_0 \\ \hat{\beta}_1 \end{pmatrix}$, using (8.39).

(d) Let $\bar{y}$ be the mean of y and $\bar{x}$ be the mean of x, and show that

$$\hat{\beta}_1 = \frac{\sum_{i=1}^{n}(x_i - \bar{x})(y_i - \bar{y})}{\sum_{i=1}^{n}(x_i - \bar{x})^2} \quad \text{and} \quad \hat{\beta}_0 = \bar{y} - \hat{\beta}_1 \bar{x}.$$

8.14. In a simple regression problem, $y_i = \beta_0 + \beta_1 x_i + \varepsilon_i$, $i = 1, 2, \ldots, n$, it was found that

$$\mathbf{X}^{\mathrm{T}}\mathbf{X} = \begin{pmatrix} 10.0 & 5.0 \\ 5.0 & 11.6 \end{pmatrix} \quad \text{and} \quad \mathbf{X}^{\mathrm{T}}\mathbf{y} = \begin{pmatrix} 0.36 \\ 20.00 \end{pmatrix}.$$

(a) What are n, $\bar{y}$, $\bar{x}$, $\sum_{i=1}^{n} x_i^2$, and $\sum_{i=1}^{n} x_i y_i$?

(b) Compute the least square estimates of the intercept β_0 and the slope β_1.

8.15. Consider the no-intercept simple regression model

$$y_i = \beta_1 x_i + \varepsilon_i, \; i = 1, 2, \ldots, n,$$

or, equivalently, in matrix notation $\mathbf{y} = \beta_1 \mathbf{x} + \varepsilon$.
(a) Show that the least squares estimate of β_1 is $\hat{\beta}_1 = \mathbf{x}^{\mathrm{T}}\mathbf{y}/\mathbf{x}^{\mathrm{T}}\mathbf{x}$.
(b) Use the data in Exercise 8.14 and compute the least squares estimate of β_1.

8.16. Consider the regression model (8.36). Let $\hat{\mathbf{y}}$ be the vector of fitted values as defined in (8.47) and let $\mathbf{e}$ be the vector of residuals as defined in (8.49).
(a) Show that $\mathbf{X}^{\mathrm{T}}\mathbf{e} = \mathbf{0}$; that is, $\mathbf{e}$ is orthogonal to each column of $\mathbf{X}$.
(b) Show that $\mathbf{e}^{\mathrm{T}}\hat{\mathbf{y}} = 0$; that is, $\mathbf{e}$ and $\hat{\mathbf{y}}$ are orthogonal to each other.
(c) Show that $(\mathbf{I} - \mathbf{P})\mathbf{X} = \mathbf{0}$; that is, the rows of $(\mathbf{I} - \mathbf{P})$ are orthogonal to the columns of $\mathbf{X}$.
(d) What are the orders of the null matrices in (a)–(c)?
(e) Compute the projection matrix $\mathbf{P}$ in (8.48) for the data matrix $\mathbf{X}$ in (8.42).
(f) Verify that $\mathbf{P}$ is symmetric and idempotent and that $r(\mathbf{P}) = tr(\mathbf{P}) = 2$.
(g) Using $\mathbf{X}$ in (8.42), $\hat{\mathbf{y}}$ in (8.45), $\mathbf{e}$ in (8.46), and $\mathbf{P}$ above, verify (a)–(c).

8.17. Consider fitting the linear regression model in (8.36) to the following data:

$$\mathbf{X} = \begin{pmatrix} 1 & -1 & 0 \\ 1 & 1 & 0 \\ -1 & 0 & 1 \\ -1 & 0 & 1 \end{pmatrix} \quad \text{and} \quad \mathbf{y} = \begin{pmatrix} 1 \\ 1 \\ 0 \\ 0 \end{pmatrix}.$$

(a) Compute $\mathbf{X}^T\mathbf{X}$.
(b) Partition $\mathbf{X}^T\mathbf{X}$ as in (6.8), then compute $(\mathbf{X}^T\mathbf{X})^{-1}$ using (6.9) or (6.10).
(c) Find the least squares estimate of β, $\hat{\beta}$, using (8.39).
(d) Compute the vector of fitted values $\hat{\mathbf{y}} = \mathbf{X}\hat{\beta}$.
(e) Compute the vector of residuals $\mathbf{e}$ using (8.44).
(f) Verify that $\mathbf{e}$ and $\hat{\mathbf{y}}$ are orthogonal vectors.

8.18. Consider fitting the linear regression model in (8.36) to the following data:

$$\mathbf{X} = \begin{pmatrix} 1 & -1 & 0 \\ 1 & 1 & 2 \\ -1 & 0 & -1 \\ -1 & 0 & -1 \end{pmatrix} \quad \text{and} \quad \mathbf{y} = \begin{pmatrix} 1 \\ 3 \\ -1 \\ -3 \end{pmatrix}.$$

(a) Verify that $\mathbf{X}$ is rank-deficient, hence the least squares estimate $\hat{\beta}$ is not unique.
(b) Compute $\mathbf{X}^T\mathbf{X}$ and $\mathbf{X}^T\mathbf{y}$ and find a solution to the system of normal equations in (8.38). [You may use (8.11).]
(c) Denote your solution in (b) by $\hat{\beta}_1$, then compute the vector of fitted values $\hat{\mathbf{y}}_1 = \mathbf{X}\hat{\beta}_1$.
(d) Find another solution to the normal equations and denote it by $\hat{\beta}_2$. [You may use (8.12).]
(e) Compute the vector of fitted values $\hat{\mathbf{y}}_2 = \mathbf{X}\hat{\beta}_2$.
(f) Verify that $\hat{\mathbf{y}}_1 = \hat{\mathbf{y}}_2$.
(g) Show that, in general, the vector of fitted values $\hat{\mathbf{y}} = \mathbf{X}\hat{\beta}$ is unique, hence, it does not depend on which solution $\hat{\beta}$ one uses for the normal equations.

8.19. Suppose that after the least squares estimate in Exercise 8.12(a) was computed, the data $(1 \quad 1 \quad -1 \quad -1)^T$ on a new predictor variable have become available.
(a) Use the updating formulas in Section 8.4.1 to update the estimate obtained in Exercise 8.12(a).
(b) Since the new $\mathbf{X}$ matrix is the same as $\mathbf{X}$ in Exercise 8.12(b), check that the updated estimate is the same as the estimate obtained in Exercise 8.12(b).

8.20. Suppose that after the least squares estimate in Exercise 8.12(a) was computed, a new observation $y = 1$ and its corresponding values of $\mathbf{x} = (1 \quad 1)$ have become available. Use the updating formulas in Section 8.4.2 to update the estimate obtained in Exercise 8.12(a).

9 Eigenvalues and Eigenvectors

Eigenvalues and eigenvectors are scalars and vectors, respectively, associated with square matrices that encapsulate useful information about a given matrix. Eigenvalues and eigenvectors have numerous applications in many fields of study. For example, they are of fundamental importance in statistics in general, and in regression and multivariate analysis in particular. In this chapter, we define, derive, and present some properties of the eigenvalues and eigenvectors of square matrices. We also present a related matrix decomposition known as the *singular value decomposition*. In Chapters 10 and 11 we discuss several applications of eigenvalues and eigenvectors.

9.1. Definition and Derivation

Consider the following question: Given a $k \times k$ matrix $\mathbf{A}$, can we find a scalar λ and a nonzero vector $\mathbf{x}$ such that $\mathbf{Ax} = \lambda\mathbf{x}$? This is known as the *eigenvalue* problem, and the answer is yes. In fact, there exist k (possibly nondistinct) values for λ and infinitely many vectors $\mathbf{x}$ satisfying

$$\mathbf{Ax} = \lambda\mathbf{x}. \tag{9.1}$$

Equation (9.1) has some interesting algebraic and geometric interpretations. Algebraically speaking, the eigenvalue problem can be considered a matrix reduction problem because the matrix $\mathbf{A}$ on the left-hand side of (9.1) is replaced by the scalar λ on the right-hand side. Note that $\mathbf{Ax}$ is a vector and is equal to $\lambda\mathbf{x}$; thus, because $\mathbf{Ax}$ is equal to a scalar multiple of $\mathbf{x}$, the vectors $\mathbf{Ax}$ and $\mathbf{x}$ are linearly dependent.

Let us now look at (9.1) from the geometric point of view. A vector $\mathbf{x}$ and $\lambda\mathbf{x}$ are drawn in Figure 9.1. These vectors have either the same direction (when $\lambda > 0$, as in Figure 9.1(a)) or opposite directions (when $\lambda < 0$, as in Figure 9.1(b)). We have also seen in Chapters 4 and 7 that when a vector $\mathbf{x}$ is multiplied by a matrix $\mathbf{A}$, the matrix usually rotates $\mathbf{x}$. If $\mathbf{x}$ satisfies (9.1), however, $\mathbf{A}$ does not rotate $\mathbf{x}$ but rather it stretches $\mathbf{x}$ (when $|\lambda| > 1$), shrinks $\mathbf{x}$ (when $|\lambda| < 1$), or leaves $\mathbf{x}$ unchanged (when $|\lambda| = 1$). In addition, when $\lambda < 0$, $\mathbf{A}$ reverses the direction of $\mathbf{x}$. Other interesting geometric interpretations are given in Section 10.1.

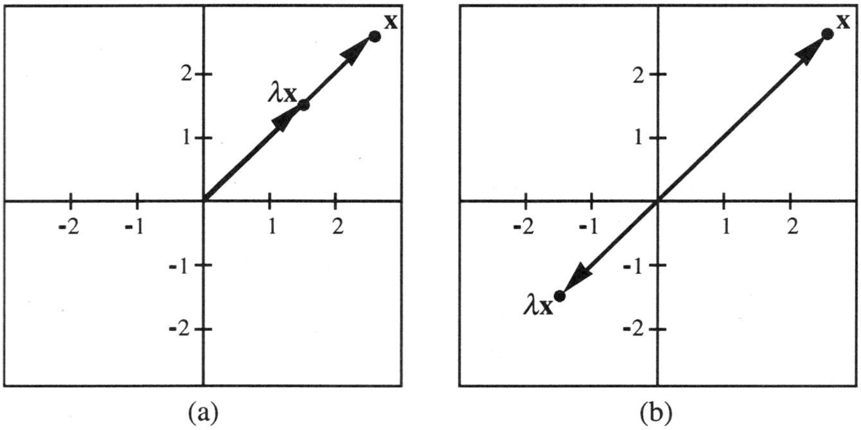

Figure 9.1. The vector $\mathbf{Ax} = \lambda\mathbf{x}$ is a scalar multiple of $\mathbf{x}$; in (a) λ is positive and in (b) λ is negative.

We now return to the question of finding λ and $\mathbf{x}$ for any given $k \times k$ matrix $\mathbf{A}$. We start by writing (9.1) as $\mathbf{Ax} - \lambda\mathbf{x} = \mathbf{0}$, then factor $\mathbf{x}$ to the right and obtain

$$(\mathbf{A} - \lambda\mathbf{I})\mathbf{x} = \mathbf{0}, \tag{9.2}$$

where $\mathbf{0}$ is the null vector. Notice that we have multiplied λ by the identity matrix, a trick which makes $\mathbf{A}$ and $\lambda\mathbf{I}$ conformable for subtraction. Notice that if λ were known, (9.2) would have the form of a homogeneous system of linear equations, as in (8.13). Recall from Section 8.1.5 that the system in (9.2) has the unique solution $\mathbf{x} = \mathbf{0}$ if and only if $(\mathbf{A} - \lambda\mathbf{I})$ is nonsingular. But $\mathbf{x} = \mathbf{0}$ is not the solution we are seeking, because $\mathbf{x}$ is required to be nonzero. It follows, then, that for $\mathbf{x}$ to be nonzero, $(\mathbf{A} - \lambda\mathbf{I})$ must be singular or, equivalently,

$$\det(\mathbf{A} - \lambda\mathbf{I}) = 0. \tag{9.3}$$

Equation (9.3) is called the *characteristic* equation. By Result 5.4(b), $\det(\mathbf{A} - \lambda\mathbf{I}) = 0$ implies that $\mathrm{r}(\mathbf{A} - \lambda\mathbf{I}) < k$. So, by Result 8.1(c), the system has infinitely many solutions.

Note that $\mathbf{A}$ is known and given, so a possible strategy for finding λ and $\mathbf{x}$ is to first solve (9.3) for λ, then substitute λ in (9.2) and solve for $\mathbf{x}$. Perhaps this is best conveyed by an example. Let

$$\mathbf{A} = \begin{pmatrix} 6 & 2 \\ 2 & 9 \end{pmatrix}; \tag{9.4}$$

then

$$\mathbf{A} - \lambda\mathbf{I} = \begin{pmatrix} 6 & 2 \\ 2 & 9 \end{pmatrix} - \lambda\begin{pmatrix} 1 & 0 \\ 0 & 1 \end{pmatrix}$$

$$= \begin{pmatrix} 6-\lambda & 2 \\ 2 & 9-\lambda \end{pmatrix}.$$

Note that λ is subtracted only from the diagonal elements of **A**, because **I** is diagonal. Thus,

$$\det(\mathbf{A} - \lambda\mathbf{I}) = (6-\lambda)(9-\lambda) - 4 = \lambda^2 - 15\,\lambda + 50 = 0. \tag{9.5}$$

Equation (9.5) is a quadratic equation, i.e., the polynomial involved is of degree 2, the order of **A**. In general, for a $k \times k$ matrix **A**, (9.3) represents a polynomial equation of degree k. From scalar algebra, a polynomial of degree k has k (possibly nondistinct) roots. These roots are called the *eigenvalues* of **A**.[1]

If the eigenvalues of **A** are not complex, we can always arrange them in order. Whenever possible, let us then arrange the eigenvalues in descending order and denote them by

$$\lambda_1 \geq \lambda_2 \geq \ldots \geq \lambda_k. \tag{9.6}$$

Thus, for example, λ_1 and λ_k are the largest and smallest eigenvalues of **A**, respectively.

There is no general equation for computing the roots of a polynomial of degree k. The computations involve many iterations and we usually rely on computers to compute the roots of polynomials. However, when $k = 2$, (9.3) is a quadratic equation, and in this case, there is a formula for computing the two roots. From scalar algebra, we know that any quadratic equation in λ can be written in the form $a\lambda^2 + b\lambda + c = 0$ and that the roots of this equation are given by the *quadratic formula*:

$$\lambda_1 = \frac{-b + \sqrt{b^2 - 4ac}}{2a} \quad \text{and} \quad \lambda_2 = \frac{-b - \sqrt{b^2 - 4ac}}{2a}. \tag{9.7}$$

Thus, for example, the roots of (9.5) are

$$\lambda_1 = \frac{-(-15) + \sqrt{(-15)^2 - 4(1)(50)}}{2(1)} = 10$$

and

$$\lambda_2 = \frac{-(-15) - \sqrt{(-15)^2 - 4(1)(50)}}{2(1)} = 5.$$

---

1 The λs are also known by many other names, such as characteristic values, eigenroots, characteristic roots, and latent roots.

Thus, the eigenvalues of **A** in (9.4) are $\lambda_1 = 10$ and $\lambda_2 = 5$. You may want to check for yourself that both of these values satisfy (9.5).

We now need to compute the eigenvectors. To do this we substitute each of the eigenvalues one at a time in (9.2) and solve for **x**. For example, for **A** in (9.4) and $\lambda_1 = 10$, (9.2) becomes

$$\begin{pmatrix} 6-\lambda_1 & 2 \\ 2 & 9-\lambda_1 \end{pmatrix} \begin{pmatrix} x_1 \\ x_2 \end{pmatrix} = \begin{pmatrix} 0 \\ 0 \end{pmatrix}$$

or

$$\begin{pmatrix} -4 & 2 \\ 2 & -1 \end{pmatrix} \begin{pmatrix} x_1 \\ x_2 \end{pmatrix} = \begin{pmatrix} 0 \\ 0 \end{pmatrix}. \tag{9.8}$$

This represents a homogeneous system of linear equations. Furthermore, the coefficient matrix has a determinant of 0 and a rank of 1. By Result 8.1(c), this system has infinitely many solutions. However, these solutions are proportional to one another; that is, if **x** is a solution to (9.2), so is $\alpha\mathbf{x}$, where α is any nonzero scalar. Thus, if we know one solution, we know them all. By Result 8.1(c), a solution can be obtained by setting some variables arbitrarily and solving for the others. For example, by performing the multiplication in (9.8), we get $-4x_1 + 2x_2 = 0$. Setting $x_1 = 1$, say, we obtain $x_2 = 2$. Therefore,

$$\mathbf{x} = \begin{pmatrix} x_1 \\ x_2 \end{pmatrix} = \begin{pmatrix} 1 \\ 2 \end{pmatrix} \tag{9.9}$$

(and any nonzero scalar multiple of **x**) is a solution. To check our calculations, we substitute this value of **x** in (9.8) and see that

$$\begin{pmatrix} -4 & 2 \\ 2 & -1 \end{pmatrix} \begin{pmatrix} 1 \\ 2 \end{pmatrix} = \begin{pmatrix} 0 \\ 0 \end{pmatrix}.$$

Therefore, the vector **x** in (9.9) is an eigenvector of the matrix **A** in (9.4) associated with $\lambda_1 = 10$, the largest eigenvalue of **A**.

As we mentioned above, in this case there are infinitely many eigenvectors associated with each of the two eigenvalues of **A**. However, if we additionally require **x** to be a normal vector, all solutions associated with a particular value of λ will be identical (up to algebraic sign), and hence the system (9.8) will have a unique (up to algebraic sign) solution. This fact follows from Result 5.5(c), because

$$\frac{\alpha\mathbf{x}}{\|\alpha\mathbf{x}\|} = \frac{\alpha\mathbf{x}}{|\alpha| \times \|\mathbf{x}\|} = \pm\frac{\mathbf{x}}{\|\mathbf{x}\|},$$

depending on the sign of α, which shows that the normalized version of $\alpha\mathbf{x}$ has either the same or opposite direction as the normalized version of **x**. Therefore, let us find one solution, normalize it, and label it $\mathbf{v}_1$. The vector $\mathbf{v}_1$ is called a *normalized eigenvector* associated with the eigenvalue λ_1. Continuing with our example,

$$\mathbf{v}_1 = \frac{\mathbf{x}}{\|\mathbf{x}\|} = \frac{1}{\sqrt{5}}\begin{pmatrix}1\\2\end{pmatrix} = \begin{pmatrix}1/\sqrt{5}\\2/\sqrt{5}\end{pmatrix}.$$

We shall refer to any eigenvalue and its associated normalized eigenvector as an *eigenpair*. Thus, $\{\lambda_1, \mathbf{v}_1\}$ are the first eigenpair of $\mathbf{A}$.

To find the second eigenpair, we perform similar calculations. We substitute $\lambda_2 = 5$ in (9.2), and obtain

$$\begin{pmatrix}1 & 2\\2 & 4\end{pmatrix}\begin{pmatrix}x_1\\x_2\end{pmatrix} = \begin{pmatrix}0\\0\end{pmatrix},$$

one solution of which is

$$\mathbf{x} = \begin{pmatrix}x_1\\x_2\end{pmatrix} = \begin{pmatrix}-2\\1\end{pmatrix}. \tag{9.10}$$

The normalized version of this solution is

$$\mathbf{v}_2 = \frac{\mathbf{x}}{\|\mathbf{x}\|} = \frac{1}{\sqrt{5}}\begin{pmatrix}-2\\1\end{pmatrix} = \begin{pmatrix}-2/\sqrt{5}\\1/\sqrt{5}\end{pmatrix},$$

which is the normalized eigenvector associated with $\lambda_2 = 5$. Thus, $\{\lambda_2, \mathbf{v}_2\}$ are the second eigenpair of $\mathbf{A}$. You may verify that $\| \mathbf{v}_1 \| = \| \mathbf{v}_2 \| = 1$ and that each of the two eigenpairs satisfies (9.1).

Although it may be a useful exercise to compute the eigenvalues and eigenvectors for a small matrix like the one in (9.4), in practice the computations are usually carried out by a computer. Hence, the computational details are not important here; what is important are the concepts and the properties that the eigenvalues and eigenvectors possess. We turn to this next.

9.2. Some Properties

Let $\Lambda = diag(\lambda_j)$ be the diagonal matrix containing the ordered eigenvalues of $\mathbf{A}$. Also, let us arrange the corresponding normalized eigenvectors in a matrix $\mathbf{V}$, say, in which the jth column is the normalized eigenvector associated with λ_j. Thus,

$$\Lambda = \begin{pmatrix}\lambda_1 & 0 & \cdots & 0\\0 & \lambda_2 & \cdots & 0\\\vdots & \vdots & \ddots & \vdots\\0 & 0 & \cdots & \lambda_k\end{pmatrix} \quad \text{and} \quad \mathbf{V} = (\mathbf{v}_1 : \mathbf{v}_2 : \ldots : \mathbf{v}_k). \tag{9.11}$$

The matrix Λ is called the *spectral* matrix and the matrix $\mathbf{V}$ is called the *normal matrix*. Thus, for the matrix $\mathbf{A}$ of (9.4) we have

$$\mathbf{A} = \begin{pmatrix} 6 & 2 \\ 2 & 9 \end{pmatrix}, \quad \Lambda = \begin{pmatrix} 10 & 0 \\ 0 & 5 \end{pmatrix}, \quad \text{and} \quad \mathbf{V} = \frac{1}{\sqrt{5}} \begin{pmatrix} 1 & -2 \\ 2 & 1 \end{pmatrix}.$$

Note that the eigenvalues arising from the characteristic equation need not be distinct. If m of the roots are the same, the common root is called an *eigenvalue of multiplicity m*. If an eigenvalue occurs more than once, the corresponding eigenvectors are not uniquely determined, even when scaled to have unit lengths. Additionally, the eigenvalues and eigenvectors of a matrix have the following properties.

Result 9.1. Let $\mathbf{A}$ be a $k \times k$ matrix whose eigenvalues and eigenvectors are given in (9.11); then:

(a) $\text{tr}(\mathbf{A}) = \lambda_1 + \lambda_2 + \cdots + \lambda_k = \sum_{j=1}^{k} \lambda_j$.

(b) $\det(\mathbf{A}) = \lambda_1 \times \lambda_2 \times \cdots \times \lambda_k = \prod_{j=1}^{k} \lambda_j$.

(c) If λ is an eigenvalue of $\mathbf{A}$, then λ^{-1} is an eigenvalue of $\mathbf{A}^{-1}$, for any nonsingular $\mathbf{A}$.

(d) The eigenvalues of any orthogonal matrix are either 1 or -1.

(e) If λ is an eigenvalue of $\mathbf{A}$, then $\lambda \pm 1$ is an eigenvalue of $\mathbf{A} \pm \mathbf{I}$, where $\mathbf{I}$ is the identity matrix.

(f) The matrices **ABC**, **BCA**, and **CAB** have the same set of nonzero eigenvalues, provided that the products exist.

(g) The eigenvalues of $\mathbf{BAB}^{-1}$ and $\mathbf{A}$ are equal for any nonsingular $k \times k$ matrix $\mathbf{B}$.

In words, Parts (a) and (b) indicate that the sum of the eigenvalues is equal to the sum of the diagonal elements of $\mathbf{A}$ and that the product (denoted by Π) of the eigenvalues is equal to the determinant of $\mathbf{A}$. Thus, Result 9.1(b) provides a general (though not necessarily computationally efficient) definition of the determinant. Parts (c)–(e) are self-explanatory. Part (f) states that the nonzero eigenvalues of a matrix product are unchanged by a conformable cyclic permutation of the matrices. This property provides a simple proof of Part (g). The product $\mathbf{BAB}^{-1}$ has the same nonzero eigenvalues as the product $\mathbf{B}^{-1}\mathbf{BA} = \mathbf{A}$.

Result 9.2. Let $\mathbf{A}$ be a $k \times k$ matrix whose eigenvalues and eigenvectors are given in (9.11); then:

(a) $\mathbf{A}$ can be written as $\mathbf{A} = \mathbf{V}\Lambda\mathbf{V}^{-1}$ if and only if the set of k eigenvectors is linearly independent, that is, if and only if $\mathbf{V}$ is nonsingular.

(b) $\mathbf{A}^n = \mathbf{V}\Lambda^n\mathbf{V}^{-1}$, for any scalar n (usually integer) provided that Λ^n exists, if and only if $\mathbf{V}$ is nonsingular.

Part (a) indicates that if $\mathbf{V}$ is nonsingular, then $\mathbf{A}$ can be decomposed into the product of three matrices. Because one of the three matrices is diagonal, the decomposition is also known as the *diagonalization* of $\mathbf{A}$, and $\mathbf{A}$ is said to be *diagonalizable*. Note that not all square matrices are diagonalizable; a $k \times k$ matrix can be diagonalized only if it has a set of k linearly independent eigenvectors. For example, the reader may verify that

$$\mathbf{A} = \begin{pmatrix} 2 & 1 \\ 0 & 2 \end{pmatrix}$$

has eigenvalues 2 and 2 but only one linearly independent eigenvector; therefore, $\mathbf{A}$ is not diagonalizable.

Part (b) provides a neat and efficient way to multiply a matrix by itself n times, especially when n is large. Note that Λ^n is simply a diagonal matrix whose jth diagonal element is λ_j^n. In Section 11.4, we shall encounter a practical example in which we make use of Result 9.2(b).

In statistical applications, most if not all matrices for which we need to compute the eigenvalues and eigenvectors are covariance or correlation matrices (see Section 8.2). These matrices are symmetric. The eigenvalues and eigenvectors of symmetric matrices have the following additional properties.

Result 9.3. Let $\mathbf{A}$ be a $k \times k$ symmetric matrix whose eigenvalues and eigenvectors are given in (9.11), let $\mathbf{I}_k$ be the identity matrix of order $k \times k$, and let $\mathbf{x}$ be a vector of order $k \times 1$. Then:

(a) The eigenvalues of $\mathbf{A}$ are all real.

(b) If an eigenvalue of $\mathbf{A}$, λ say, is of multiplicity m, then there exist m mutually orthogonal eigenvectors corresponding to this value of λ.

(c) $\mathbf{V}$ is an orthogonal matrix and $\mathbf{V}^{-1} = \mathbf{V}^{\mathrm{T}}$ or, equivalently, $\mathbf{V}\mathbf{V}^{\mathrm{T}} = \mathbf{V}^{\mathrm{T}}\mathbf{V} = \mathbf{I}_k$.

(d) $\mathbf{A}$ is diagonalizable and can be expressed as $\mathbf{A} = \mathbf{V}\Lambda\mathbf{V}^{\mathrm{T}}$.

(e) $\mathbf{A} = \sum_{j=1}^{k} \lambda_j \mathbf{v}_j \mathbf{v}_j^{\mathrm{T}}$.

(f) $\sum_{j=1}^{k} \mathbf{v}_j \mathbf{v}_j^{\mathrm{T}} = \mathbf{V}^{\mathrm{T}}\mathbf{V} = \mathbf{I}_k$.

(g) $\mathbf{A}^n = \mathbf{V}\Lambda^n\mathbf{V}^{\mathrm{T}}$, for any scalar n.

(h) $\max_{\mathbf{x} \neq \mathbf{0}} \dfrac{\mathbf{x}^{\mathrm{T}}\mathbf{A}\mathbf{x}}{\mathbf{x}^{\mathrm{T}}\mathbf{x}} = \lambda_1$, where λ_1 is the largest eigenvalue of $\mathbf{A}$.

(i) $\min_{\mathbf{x} \neq \mathbf{0}} \dfrac{\mathbf{x}^{\mathrm{T}}\mathbf{A}\mathbf{x}}{\mathbf{x}^{\mathrm{T}}\mathbf{x}} = \lambda_k$, where λ_k is the smallest eigenvalue of $\mathbf{A}$.

(j) The number of nonzero eigenvalues of $\mathbf{A}$ is equal to the rank of $\mathbf{A}$.

A matrix can have complex eigenvalues even if all its elements are real numbers. If $\mathbf{A}$ is symmetric, however, Part (a) guarantees that all eigenvalues of $\mathbf{A}$ are real numbers. We shall see an example of Part (b) in Section 10.1. Part (c) indicates that the inverse of $\mathbf{V}$ is equal to its transpose. Part (d) is known as the *spectral decomposition* of $\mathbf{A}$.

It is a special case of Result 9.2(a). It implies that all symmetric matrices are diagonalizable. It also indicates that **A** can be decomposed into the product of three matrices.

Note that $\mathbf{v}_j\mathbf{v}_j^T$ is a matrix of order $k \times k$; hence Part (e) indicates that **A** can be decomposed into a sum of k matrices. An interesting application of this result is obtained when some of the eigenvalues of **A** are 0. For example, if only the first $h < k$ eigenvalues are nonzero, then **A** can be written as

$$\mathbf{A} = \sum_{j=1}^{h} \lambda_j \mathbf{v}_j \mathbf{v}_j^T,$$

which implies that **A** can be constructed using only the first h eigenpairs.

Part (f) indicates that $\mathbf{V}^T\mathbf{V}$ can be decomposed into a sum of k matrices. Part (g) is a special case of Result 9.2(b). Two important special cases come out of Result 9.3(g): when n is -1 or $1/2$. When $n = -1$, we have $\mathbf{A}^{-1} = \mathbf{V}\Lambda^{-1}\mathbf{V}^{-1}$, which is easily verified as follows. Using Result 9.3(c),

$$\mathbf{A}\mathbf{A}^{-1} = \mathbf{V}\Lambda\mathbf{V}^{-1}\mathbf{V}\Lambda^{-1}\mathbf{V}^{-1} = \mathbf{V}\Lambda\,\Lambda^{-1}\mathbf{V}^{-1} = \mathbf{V}\mathbf{V}^{-1} = \mathbf{I}.$$

When $n = 1/2$ (and all eigenvalues are nonnegative so that $\Lambda^{1/2}$ exists), we have $\mathbf{A}^{1/2} = \mathbf{V}\Lambda^{1/2}\mathbf{V}^{-1}$, which is known as a *symmetric square root* of **A**. The reader can verify that $\mathbf{A}^{1/2}$ is symmetric and $\mathbf{A}^{1/2}\mathbf{A}^{1/2} = \mathbf{A}$.

Parts (h) and (i) indicate that for all vectors **x**, the maximum and minimum values of the ratio $\mathbf{x}^T\mathbf{A}\mathbf{x} / \mathbf{x}^T\mathbf{x}$ are the largest and smallest eigenvalues of **A**, respectively.

Finally, Part (j) is in fact a special case of a more general result which states that, for any square matrix **A** (not necessarily symmetric), the number of nonzero eigenvalues of **A** never exceeds r(**A**), but if **A** is symmetric, this number is exactly equal to r(**A**).

Using Results 9.1 and 9.3 to check our calculations in the example from Section 9.1, we have $\lambda_1 + \lambda_2 = 15 = \text{tr}(\mathbf{A})$, and $\lambda_1\lambda_2 = 50 = \det(\mathbf{A})$, as would be expected. We also see that

$$\lambda_1\mathbf{v}_1\mathbf{v}_1^T = 10\begin{pmatrix} 1/\sqrt{5} \\ 2/\sqrt{5} \end{pmatrix}\begin{pmatrix} 1/\sqrt{5} & 2/\sqrt{5} \end{pmatrix} = \begin{pmatrix} 2 & 4 \\ 4 & 8 \end{pmatrix}$$

and

$$\lambda_2\mathbf{v}_2\mathbf{v}_2^T = 5\begin{pmatrix} -2/\sqrt{5} \\ 1/\sqrt{5} \end{pmatrix}\begin{pmatrix} -2/\sqrt{5} & 1/\sqrt{5} \end{pmatrix} = \begin{pmatrix} 4 & -2 \\ -2 & 1 \end{pmatrix}.$$

Thus,

$$\lambda_1\mathbf{v}_1\mathbf{v}_1^T + \lambda_2\mathbf{v}_2\mathbf{v}_2^T = \begin{pmatrix} 2 & 4 \\ 4 & 8 \end{pmatrix} + \begin{pmatrix} 4 & -2 \\ -2 & 1 \end{pmatrix} = \begin{pmatrix} 6 & 2 \\ 2 & 9 \end{pmatrix} = \mathbf{A},$$

as would be expected. The reader can easily verify the remaining properties.

We end this section with three more properties of eigenvalues.

Result 9.4. All eigenvalues of variance-covariance and correlation matrices are nonnegative.[2]

Result 9.5. Let **P** be a symmetric and idempotent matrix[3] of order $n \times n$ and of rank k; then k of the eigenvalues of **P** are equal to 1 and the remaining $n - k$ are equal to 0.

Recall from Section 8.2 that the variance-covariance matrix **S** and the correlation matrix **R** are symmetric. Therefore, the eigenvalues and eigenvectors of **S**, **R**, and **P** above possess all of the properties listed in Result 9.3 as well.

Result 9.6. Eigenvalues of a Kronecker product. Let **A** be of order $n \times n$ and **B** be of order $m \times m$. Then:

(a) The nm eigenvalues of $\mathbf{A}\otimes\mathbf{B}$ are the products of the n eigenvalues of **A** with the m eigenvalues of **B**.

(b) $\det(\mathbf{A}\otimes\mathbf{B}) = \left(\prod_{i=1}^{n} \lambda_i\right)^m \left(\prod_{j=1}^{m} \gamma_j\right)^n.$

This result perhaps needs some explanation. Let $\lambda_1, \lambda_2, \ldots, \lambda_n$ be the eigenvalues of **A** and $\gamma_1, \gamma_2, \ldots, \gamma_m$ be the eigenvalues of **B**. These eigenvalues can be arranged in two vectors such as

$$\boldsymbol{\lambda} = \begin{pmatrix} \lambda_1 \\ \lambda_2 \\ \vdots \\ \lambda_n \end{pmatrix} \quad \text{and} \quad \boldsymbol{\gamma} = \begin{pmatrix} \gamma_1 \\ \gamma_2 \\ \vdots \\ \gamma_m \end{pmatrix}.$$

The product $\mathbf{A}\otimes\mathbf{B}$ is of order $nm \times nm$ and has nm eigenvalues. These eigenvalues can be organized in a matrix as follows:

$$\begin{pmatrix} \lambda_1\gamma_1 & \lambda_1\gamma_2 & \cdots & \lambda_1\gamma_m \\ \lambda_2\gamma_1 & \lambda_2\gamma_2 & \cdots & \lambda_2\gamma_m \\ \vdots & \vdots & \ddots & \vdots \\ \lambda_n\gamma_1 & \lambda_n\gamma_2 & \cdots & \lambda_n\gamma_m \end{pmatrix} = \boldsymbol{\lambda}\boldsymbol{\gamma}^{\mathrm{T}}. \tag{9.12}$$

From (2.16), it follows that $\boldsymbol{\lambda}\boldsymbol{\gamma}^{\mathrm{T}} = \boldsymbol{\lambda}\otimes\boldsymbol{\gamma}^{\mathrm{T}}$. Thus, Result 9.6(a) asserts that the eigenvalues of $\mathbf{A}\otimes\mathbf{B}$ are the elements of $\boldsymbol{\lambda}\otimes\boldsymbol{\gamma}^{\mathrm{T}}$.

For Part (b), it follows from (9.12) and Result 9.1(b) that

---

[2] Variance-covariance and correlation matrices are defined in (8.28) and (8.31), respectively.

[3] Recall from Section 2.3.2 that any matrix **P** satisfying **PP** = **P** is called an idempotent matrix.

$$\det(\mathbf{A}\otimes\mathbf{B}) = \left(\prod_{i=1}^{n} \lambda_i\right)^m \left(\prod_{j=1}^{m} \gamma_j\right)^n = (\det(\mathbf{A}))^m\,(\det(\mathbf{B}))^n.$$

The left-hand equation is Result 9.6(b) and the right-hand equation is Result 5.4(h).

*9.3. Singular Value Decomposition

Eigenvalues and eigenvectors are defined only for square matrices and, by the spectral decomposition theorem, any symmetric matrix can be decomposed into the product of three matrices as given in Result 9.3(d). There exists a related decomposition for general matrices, square or rectangular. This decomposition is defined in the following result.

Result 9.7. Let $\mathbf{X}$ be an $n \times k$ matrix with $n \geq k$. Then $\mathbf{X}$ can be expressed as

$$\mathbf{X} = \mathbf{U}\Delta\mathbf{V}^{\mathrm{T}}, \tag{9.13}$$

where $\mathbf{U}$ is an $n \times k$ matrix with orthogonal columns ($\mathbf{U}^{\mathrm{T}}\mathbf{U} = \mathbf{I}_k$), Δ is a $k \times k$ diagonal matrix with nonnegative elements, and $\mathbf{V}$ is a $k \times k$ orthogonal matrix.

Equation (9.13) indicates that $\mathbf{X}$ can be expressed as a product of three matrices. This decomposition is known as the *singular value decomposition* of $\mathbf{X}$.[4] The diagonal elements of Δ are known as the *singular values* of $\mathbf{X}$. The singular values are the positive square roots of the eigenvalues of $\mathbf{X}^{\mathrm{T}}\mathbf{X}$. To see this we write

$$\mathbf{X}^{\mathrm{T}}\mathbf{X} = \mathbf{V}\Delta\mathbf{U}^{\mathrm{T}}\mathbf{U}\Delta\mathbf{V}^{\mathrm{T}} = \mathbf{V}\Delta^2\mathbf{V}^{\mathrm{T}}$$

Post-multiplying both sides by $\mathbf{V}$, we obtain

$$\mathbf{X}^{\mathrm{T}}\mathbf{X}\mathbf{V} = \mathbf{V}\Delta^2\mathbf{V}^{\mathrm{T}}\mathbf{V} = \mathbf{V}\Delta^2,$$

which implies

$$\mathbf{X}^{\mathrm{T}}\mathbf{X}\mathbf{v}_j = \delta_j^2\mathbf{v}_j, \tag{9.14}$$

where $\mathbf{v}_j$ is the jth column of $\mathbf{V}$ and δ_j is the jth diagonal element of Δ. By comparing (9.14) with (9.1), we conclude that δ_j^2 is the jth eigenvalue and $\mathbf{v}_j$ is the corresponding eigenvector of $\mathbf{X}^{\mathrm{T}}\mathbf{X}$. The columns of $\mathbf{U}$ can also be shown to be eigenvectors of $\mathbf{X}\mathbf{X}^{\mathrm{T}}$. These and other properties are summarized as follows.

Result 9.8. The matrices defined in (9.13) have the following properties:
(a) If δ is a singular value of $\mathbf{X}$, then $\lambda = \delta^2$ is an eigenvalue of $\mathbf{X}^{\mathrm{T}}\mathbf{X}$, that is, the eigenvalues of $\mathbf{X}^{\mathrm{T}}\mathbf{X}$ are the squares of the singular values of $\mathbf{X}$.

---

4 We should note here that there are other forms of the singular value decomposition of $\mathbf{X}$ that may be useful in other contexts. All variations have the same form as in (9.13), but they differ in the way the matrices involved are defined. See, e.g., Belsley (1991).

(b) The squares of the diagonal elements of Δ are the eigenvalues of $\mathbf{X}^T\mathbf{X}$, and the matrix $\mathbf{V}$ consists of the corresponding eigenvectors.
(c) If there are p singular values equal to 0, then $\mathbf{X}\mathbf{V}_0 = \mathbf{0}$, where $\mathbf{0}$ is the null matrix and $\mathbf{V}_0$ is the matrix containing the corresponding p columns of $\mathbf{V}$.
(d) The matrix $\mathbf{U}$ consists of a subset of the eigenvectors of $\mathbf{X}\mathbf{X}^T$. It includes the eigenvectors corresponding to the nonzero eigenvalues of $\mathbf{X}\mathbf{X}^T$.
(e) The nonzero eigenvalues of $\mathbf{X}\mathbf{X}^T$ are the same as those of $\mathbf{X}^T\mathbf{X}$.
(f) The matrices $\mathbf{X}^T\mathbf{X}$ and $\mathbf{X}\mathbf{X}^T$ are symmetric and their eigenvalues are nonnegative.[5]
(g) $r(\mathbf{X})$ is equal to the number of nonzero singular values of $\mathbf{X}$.

In Chapter 11 we shall give examples illustrating the usefulness of the singular values in defining the matrix norms and in diagnosing a difficult and frequently occurring problem in statistical analysis of data, the *collinearity* problem.

Exercises

9.1. Find and compare the eigenvalues and eigenvectors for each of the matrices below.

(a) $\begin{pmatrix} 2 & 1 \\ 1 & 2 \end{pmatrix}$ (b) $\begin{pmatrix} 4 & 2 \\ 2 & 4 \end{pmatrix}$ (c) $\begin{pmatrix} 8 & 4 \\ 4 & 8 \end{pmatrix}$

9.2. Let $\mathbf{A} = \begin{pmatrix} -4 & 2 \\ 2 & -1 \end{pmatrix}$ and $\mathbf{c} = \begin{pmatrix} c_1 \\ c_2 \end{pmatrix}$.
(a) Find a g-inverse $\mathbf{A}^-$ of $\mathbf{A}$.
(b) Using $\mathbf{A}^-$ and (8.12), find the general solution of the linear system in (9.8) in terms of a vector $\mathbf{c}$ and notice that it does not depend on c_1.
(c) Using the general solution obtained in (b), show that the vector $\mathbf{x}$ in (9.9) or any nonzero scalar multiple of it is a solution to the system in (9.8), as claimed below (9.9).

9.3. Let $\mathbf{A} = \begin{pmatrix} 2 & 4 \\ 4 & 2 \end{pmatrix}$, $\mathbf{B} = \begin{pmatrix} 3 & 4 \\ 4 & -3 \end{pmatrix}$, $\mathbf{C} = \begin{pmatrix} 1 & 2 \\ 5 & 4 \end{pmatrix}$, and $\mathbf{D} = \begin{pmatrix} 3 & -2 \\ 2 & -2 \end{pmatrix}$.
(a) Find the eigenvalues and eigenvectors for $\mathbf{A}$, $\mathbf{B}$, $\mathbf{C}$, and $\mathbf{D}$.
(b) Find the eigenvalues and eigenvectors for $\mathbf{A}^{-1}$, $\mathbf{B}^{-1}$, $\mathbf{C}^{-1}$, and $\mathbf{D}^{-1}$.
(c) Use $\mathbf{A}$, $\mathbf{B}$, and $\mathbf{C}$ to verify Result 9.1(e)–(g).
(d) Use any one of the above matrices to verify Result 9.2(a).
(e) Compute $\det(\mathbf{A}\otimes\mathbf{B})$ using Result 9.6(b).
(f) Compute $\mathrm{tr}(\mathbf{A}\otimes\mathbf{B})$.

9.4. Find the eigenvalues for each of the following matrices:

---

5 We shall see in Section 11.3 that matrices with nonnegative eigenvalues are called *positive semi-definite* matrices.

$$\mathbf{A} = \begin{pmatrix} 2 & 0 & 0 \\ 0 & 4 & 0 \\ 0 & 0 & 1 \end{pmatrix}, \quad \mathbf{B} = \begin{pmatrix} 2 & 4 & 1 \\ 0 & 2 & 2 \\ 0 & 0 & 1 \end{pmatrix}, \quad \mathbf{C} = \begin{pmatrix} 2 & 0 & 0 \\ 0 & 4 & 0 \\ 1 & 0 & 1 \end{pmatrix}.$$

9.5. Let $\mathbf{A} = \begin{pmatrix} a & 6 \\ 6 & b \end{pmatrix}$. Find a and b such that the eigenvalues of **A** are 9 and –3.

9.6. Determine whether each of the following matrices is diagonalizable.

$$\mathbf{A} = \begin{pmatrix} 2 & 1 \\ 1 & 2 \end{pmatrix}, \quad \mathbf{B} = \begin{pmatrix} 3 & 1 \\ 0 & 1 \end{pmatrix}, \quad \mathbf{C} = \begin{pmatrix} 1 & 1 \\ 1 & 3 \end{pmatrix}, \quad \mathbf{D} = \begin{pmatrix} 1 & 2 \\ 0 & 3 \end{pmatrix}.$$

9.7. Show that $\mathbf{A} = \begin{pmatrix} 2 & c \\ 0 & 2 \end{pmatrix}$ is not diagonalizable for any value of c.

9.8. The eigenvalues of a matrix **A** are 5, 2, 1, and 0. Determine whether each of the following statements is true or false.
(a) **A** is of order 4×4. (b) $\text{tr}(\mathbf{A}) = 0$. (c) $\det(\mathbf{A}) = 8$.
(d) The eigenvalues of $\mathbf{A}^{-1}$ are 1, 0.5, 0.2, and 0.
(e) The eigenvalues of $\mathbf{A}^{\mathrm{T}}$ are 5, 2, 1, and 0.
(f) The eigenvalues of $5\mathbf{A}^{-1}$ are 25, 10, 5, and 0.
(g) The matrix **A** is singular.

9.9. Let $\mathbf{A} = \begin{pmatrix} 0.5 & 0.4 \\ 0.5 & 0.6 \end{pmatrix}$. Compute $\mathbf{A}^n$, where:
(a) $n = -0.5$. (b) $n = 0.5$. (c) $n = 2$. (d) $n = \infty$.

9.10. Find the matrix **A** whose eigenvalues are 4 and 1 and the corresponding eigenvectors are

$$\mathbf{v}_1 = \begin{pmatrix} 1 \\ 2 \end{pmatrix} \quad \text{and} \quad \mathbf{v}_2 = \begin{pmatrix} -2 \\ 1 \end{pmatrix}.$$

9.11. Let $\mathbf{A} = \begin{pmatrix} 5 & 4 \\ 5 & 6 \end{pmatrix}$.

(a) Show that the characteristic equation of **A** as a function of λ is

$$f(\lambda) = \det(\mathbf{A} - \lambda\mathbf{I}) = \lambda^2 - 11\lambda + 10 = 0.$$

(b) An interesting result, known as the *Cayley-Hamilton Theorem*, states that a matrix satisfies its own characteristic equation; that is, the characteristic equation holds when λ is replaced by **A**. Verify this result using the matrix **A** above; that is, verify that $f(\mathbf{A}) = 0$.

(c) Verify the Cayley-Hamilton Theorem using **A** in (9.4) and the characteristic equation in (9.5).

9.12. Let **A** be a matrix with eigenvalues a, b, and c for which the corresponding eigenvectors are

$$\frac{1}{\sqrt{2}}\begin{pmatrix}1\\0\\1\end{pmatrix},\quad \frac{1}{\sqrt{2}}\begin{pmatrix}1\\0\\-1\end{pmatrix},\quad \begin{pmatrix}0\\1\\0\end{pmatrix}.$$

(a) Show that $\mathbf{A} = \frac{1}{2}\begin{pmatrix}a+b & 0 & a-b\\ 0 & 2c & 0\\ a-b & 0 & a+b\end{pmatrix}$. [You may use Result 9.3(d).]

(b) Use **A** and the above eigenvalues and eigenvectors to verify Parts (c), (e), and (f) of Result 9.3.
(c) Compute $\mathbf{A}^2$ first by multiplying **A** by itself, then by using Result 9.3(g).
(d) Compute $\mathbf{A}^{-1}$ using Result 9.3(g).
(e) What values of a, b, and c would make **A** singular?
(f) Assign some values to a, b, and c so that **A** has an eigenvalue of multiplicity 2.
(g) Assign some values to a, b, and c so that **A** has an eigenvalue of multiplicity 3.

9.13. Let $\mathbf{X} = \begin{pmatrix}1 & -1\\ 0 & 0\\ 1 & 1\end{pmatrix}$, $\mathbf{Y} = \begin{pmatrix}1 & 1\\ 0 & 1\\ 1 & 0\end{pmatrix}$, and $\mathbf{Z} = \begin{pmatrix}1 & -1 & 1\\ 1 & 1 & 1\\ -1 & 0 & 1\\ -1 & 0 & 1\end{pmatrix}$.

(a) Compute the eigenvalues and eigenvectors of $\mathbf{X}^T\mathbf{X}$, $\mathbf{Y}^T\mathbf{Y}$, and $\mathbf{Z}^T\mathbf{Z}$.
(b) Compute the singular values for each of **X**, **Y**, and **Z**.
(c) Compute the singular value decomposition for each of the matrices **X**, **Y**, and **Z**. [Hint: Use your answers in (a) and (b) to compute Δ and **V**, then post-multiply both sides of (9.13) by $\mathbf{V}\Delta^{-1}$ to obtain **U**.]
(d) What are the eigenvalues of $\mathbf{X}\mathbf{X}^T$, $\mathbf{Y}\mathbf{Y}^T$, and $\mathbf{Z}\mathbf{Z}^T$?
(e) What are the eigenvectors corresponding to the nonzero eigenvalues of $\mathbf{X}\mathbf{X}^T$, $\mathbf{Y}\mathbf{Y}^T$, and $\mathbf{Z}\mathbf{Z}^T$?
(f) Compute $\det(\mathbf{X}\mathbf{X}^T)$, $\det(\mathbf{Y}\mathbf{Y}^T)$, and $\det(\mathbf{Z}\mathbf{Z}^T)$.
(g) Compute $\mathrm{tr}(\mathbf{X}\mathbf{X}^T)$, $\mathrm{tr}(\mathbf{Y}\mathbf{Y}^T)$, and $\mathrm{tr}(\mathbf{Z}\mathbf{Z}^T)$.

9.14. Let **A** be a diagonalizable matrix with nonnegative eigenvalues and **B** be a nonsingular matrix of the same order.
(a) Show that $\mathbf{A}^{1/2}$ exists.
(b) Show that the square root of $\mathbf{B}\mathbf{A}\mathbf{B}^{-1}$ is $\mathbf{B}\mathbf{A}^{1/2}\mathbf{B}^{-1}$.

9.15. Show that the singular values of any real symmetric matrix are the same as its eigenvalues, and use the matrix **A** in Exercise 9.3 to verify this result.

9.16. Given that **A** is diagonalizable, show that:
(a) The eigenvalues of $\mathbf{A}^2$ are the squares of the eigenvalues of **A**.
(b) **A** and $\mathbf{A}^2$ have the same set of eigenvectors.

9.17. When two square matrices have the same set of eigenvalues (but not necessarily the same set of eigenvectors), they are said to be *similar*.
(a) Determine which pairs of the matrices in Exercise 9.6 are similar.
(b) For any two similar matrices **A** and **B**, show that $\det(\mathbf{A}) = \det(\mathbf{B})$.

9.18. A definition of similar matrices equivalent to that given in Exercise 9.17 is as follows: Two square matrices **A** and **B** are said to be similar if there exists a nonsingular matrix **V** such that $\mathbf{V}^{-1}\mathbf{A}\mathbf{V} = \mathbf{B}$.
(a) For every pair of similar matrices **A** and **B** in Exercise 9.6, find a matrix **V** such that $\mathbf{V}^{-1}\mathbf{A}\mathbf{V} = \mathbf{B}$.
(b) Show that if **A** is similar to **B** and **B** is similar to **C**, then **A** is similar to **C**.

10 Ellipsoids and Distances

In this chapter we cover two related topics: ellipsoids and distances. Section 10.1, in which we discuss the general equation of ellipsoids, serves two purposes. First, it illustrates one of many applications of eigenvalues, a topic discussed in Chapter 9. Second, it helps us understand how to measure distances in a multidimensional space, a topic presented in Section 10.2. In Section 10.3, we find the minimum-volume ellipsoid that contains a scatter of points in a space.

10.1. Ellipsoids and Spheres

We know from elementary algebra that

$$x^2 + y^2 = r^2 \tag{10.1}$$

is an equation of a *circle* centered at the origin with a *radius* equal to $r > 0$. If the circle is centered at some other point (a, b), then (10.1) becomes

$$(x - a)^2 + (y - b)^2 = r^2. \tag{10.2}$$

By dividing both sides of (10.2) by r^2, it becomes

$$\frac{(x-a)^2}{r^2} + \frac{(y-b)^2}{r^2} = 1. \tag{10.3}$$

For example, the equation of a circle centered at the origin with a radius of $r = 5$ is

$$\frac{x^2}{25} + \frac{y^2}{25} = 1. \tag{10.4}$$

This circle is shown in Figure 10.1.

Now, let us modify (10.3) a little and write

$$\frac{(x-a)^2}{u^2} + \frac{(y-b)^2}{v^2} = d^2, \tag{10.5}$$

where u, v, and d are positive scalars. Clearly, when $u^2 = v^2$, (10.5) reduces to (10.3). (Note that in this case we can divide both sides of (10.5) by d^2 so that the right-hand side becomes 1.) For example, setting $a = b = 0$, $u = 3$, $v = 2$, and $d = 4$, (10.5) becomes

$$\frac{x^2}{9} + \frac{y^2}{4} = 16 \tag{10.6}$$

or

$$\frac{x^2}{144} + \frac{y^2}{64} = 1. \tag{10.7}$$

As shown in Figure 10.2, the graph of (10.6) and (10.7) is an *ellipse* centered at the origin.

An ellipse has two *principal axes*: one is called the *major* (or *first principal*) *axis* and the other is called the *minor* (or *second principal*) *axis*. The major axis is the longest straight line passing through the center and connecting two points on the circumference of the ellipse. Similarly, the minor axis is the shortest straight line passing through the center and connecting two points on the circumference of the ellipse. The two principal axes of any ellipse are perpendicular, and, in the notation of (10.5), the lengths of the two axes are $2du$ and $2dv$. For example, the length of the major axis of the ellipse in Figure 10.2 is $2du = 24$ and the length of its minor axis is $2dv = 16$.

The ellipse represented by (10.6) or its equivalent (10.7) is said to be in the *standard position* because it is centered at the origin and its axes coincide with the horizontal and vertical axes (see Figure 10.2). In contrast, the ellipse graphed in Figure 10.3 is not in the standard position. What, then, is the equation of a general ellipse such as the one shown in Figure 10.3?

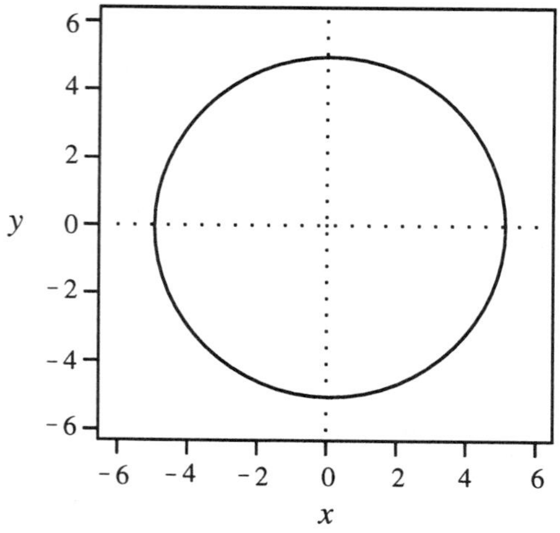

Figure 10.1. The graph of the circle given by (10.4).

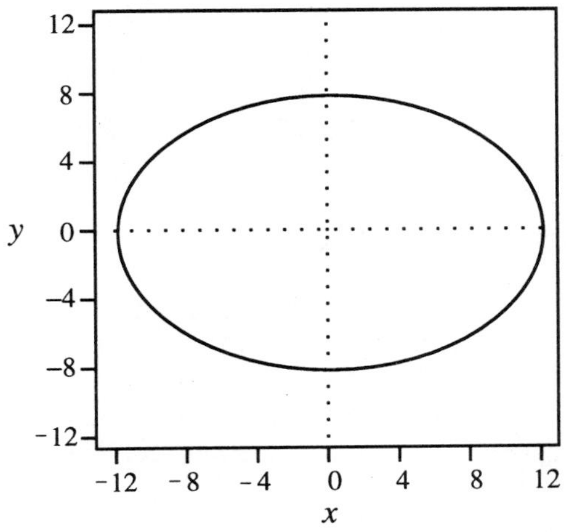

Figure 10.2. The graph of (10.6) is an ellipse in the standard position.

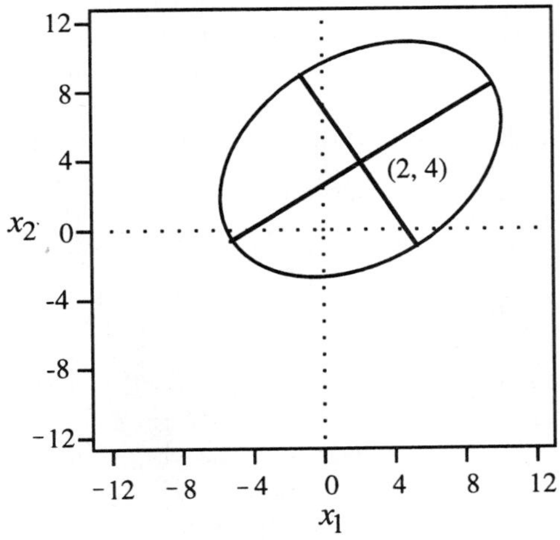

Figure 10.3. An example of a general ellipse.

To write an equation for a general ellipse, let us first start with an equation of an ellipse and try to write it in matrix notation. To simplify the notation, we replace the coordinate axes (x, y) by (x_1, x_2). This change of notation also facilitates the extension from a two-dimensional ellipse to a k-dimensional *ellipsoid*, discussed later in this section. To write (10.6), for example, in matrix notation, let

$$\mathbf{x} = \begin{pmatrix} x_1 \\ x_2 \end{pmatrix}, \quad \mathbf{c} = \begin{pmatrix} 0 \\ 0 \end{pmatrix}, \quad \mathbf{S} = \begin{pmatrix} 9 & 0 \\ 0 & 4 \end{pmatrix}, \quad \text{and} \quad d = 4; \tag{10.8}$$

then the reader can verify that (10.6) is the same as

$$(\mathbf{x} - \mathbf{c})^{\mathrm{T}}\mathbf{S}^{-1}(\mathbf{x} - \mathbf{c}) = d^2. \tag{10.9}$$

Thus, (10.9) is the equation of the ellipse graphed in Figure 10.2. By dividing both sides of (10.9) by d^2, we obtain

$$(\mathbf{x} - \mathbf{c})^{\mathrm{T}}(d^2\mathbf{S})^{-1}(\mathbf{x} - \mathbf{c}) = 1. \tag{10.10}$$

The vector **c** (viewed as a point in a space) represents the center of the ellipse. The matrix **S** controls two aspects of the ellipse: its *orientation* and *shape*. By orientation, we mean the direction of its axes and by shape we mean the relative lengths of the axes. For a fixed **S**, the scalar d determines the absolute lengths of the axes. To illustrate, suppose we replace **c**, **S**, and d in (10.8) by

$$\mathbf{c} = \begin{pmatrix} 2 \\ 4 \end{pmatrix}, \qquad \mathbf{S} = \begin{pmatrix} 7 & \sqrt{3} \\ \sqrt{3} & 5 \end{pmatrix}, \qquad \text{and} \qquad d = 3. \tag{10.11}$$

The graph of this ellipse is shown in Figure 10.3. Comparing this ellipse with the one in Figure 10.2, we notice the following changes. First, the center of the ellipse moved from the origin to the point (2, 4) as a result of changes in **c**. Second, both the orientation and shape of the ellipse have changed as a result of changes in **S**.

To further illustrate the effects of the off-diagonal elements of **S**, let us replace **S** in (10.11) by

$$\mathbf{S} = \begin{pmatrix} 7 & -\sqrt{3} \\ -\sqrt{3} & 5 \end{pmatrix}. \tag{10.12}$$

The new ellipse is drawn in Figure 10.4. The two ellipses in Figures 10.3 and 10.4 have the same center and shape but different orientations. So, generally speaking, when the off-diagonal elements are positive, the major axis has a positive slope (see Figure 10.3); when they are negative, the major axis has a negative slope (see Figure 10.4); and when they are 0, the major axis has a zero slope (see Figure 10.2).

Finally, to illustrate the effect of d, let us keep c and **S** in (10.11) fixed but change d to 2. The new ellipse is drawn in Figure 10.5. As can be seen, the two ellipses in Figures 10.3 and 10.5 have the same center, orientation, and shape; but they differ in the lengths of their axes. Thus, for fixed **S**, d determines the area of an ellipse; the larger d is, the larger the area.

Now, what would happen if we go beyond a two-dimensional space? For example, in a three-dimensional space, a circle becomes a *sphere* (or a *ball*)[1] and an ellipse becomes an *ellipsoid*. Thus, the sphere is a special case of an ellipsoid in which the three principal axes are equal in length. A football is an example of a three-dimensional ellipsoid, and a soccer ball is an example of a three-dimensional sphere.

---

1 A sphere of four or more dimensions is sometimes referred to as a *hyper-sphere*.

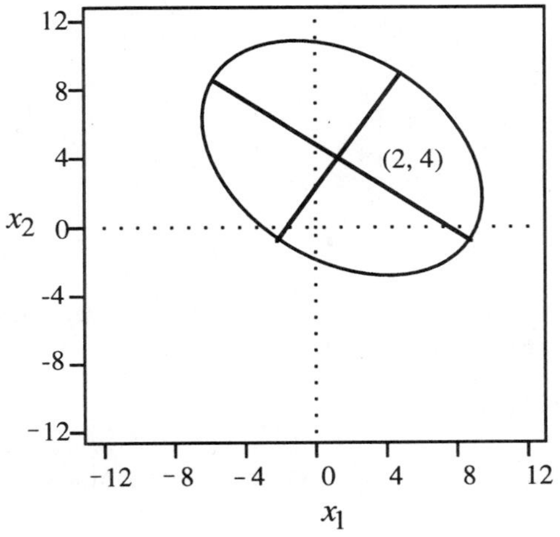

Figure 10.4. A graph of the ellipse specified by **c** and d in (10.11) and **S** in (10.12).

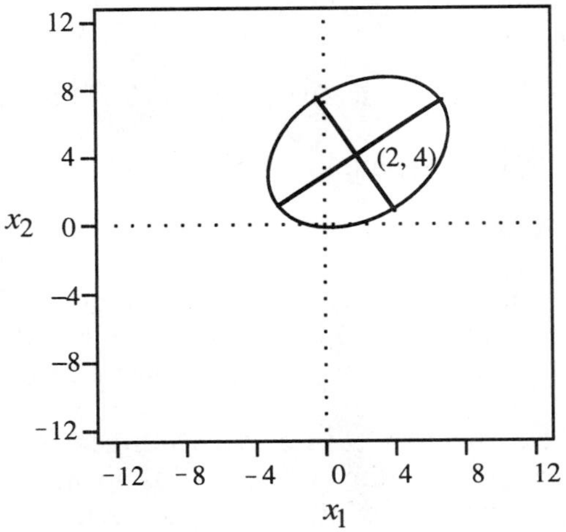

Figure 10.5. A graph of the ellipse specified by **c** and **S** in (10.11) and $d = 2$.

Without matrix algebra, it is very difficult to write the equation of a general k-dimensional ellipsoid, even when k is as small as 3 or 4. As k gets larger, the task becomes virtually impossible. With matrix algebra, it is easy to write an equation for a general multidimensional ellipsoid. In fact, (10.9) becomes an equation for a general multidimensional ellipsoid if we make appropriate changes in the dimensions of **x**, **c**, and **S**. Thus, letting

$$\mathbf{x} = \begin{pmatrix} x_1 \\ x_2 \\ \vdots \\ x_k \end{pmatrix}, \quad \mathbf{c} = \begin{pmatrix} c_1 \\ c_2 \\ \vdots \\ c_k \end{pmatrix}, \quad \text{and} \quad \mathbf{S}_{k \times k} = (s_{ij}), \tag{10.13}$$

we have the following result.

Result 10.1. Any k-dimensional ellipsoid in a k-dimensional space can be represented by an equation of the form

$$(\mathbf{x} - \mathbf{c})^{\mathrm{T}}\mathbf{S}^{-1}(\mathbf{x} - \mathbf{c}) = d^2, \tag{10.14}$$

as long as $\mathbf{S}$ is a symmetric matrix with positive eigenvalues.[2]

Several special cases can be obtained from (10.14). First, for fixed d, when $\mathbf{S}$ is diagonal with equal diagonal elements, then (10.14) is an equation of a k-dimensional sphere with radius equal to $d\sqrt{a}$, where a is the common diagonal element. For example, when $\mathbf{S} = \mathbf{I}$, (10.14) reduces to

$$(\mathbf{x} - \mathbf{c})^{\mathrm{T}}(\mathbf{x} - \mathbf{c}) = d^2, \tag{10.15}$$

which implies that $d = \|\mathbf{x} - \mathbf{c}\|$. Thus, for fixed d, (10.15) is an equation of a sphere centered at $\mathbf{c}$ with radius equal to $d\sqrt{1} = d$.

Second, when $\mathbf{c} = \mathbf{0}$, (10.14) simplifies to

$$\mathbf{x}^{\mathrm{T}}\mathbf{S}^{-1}\mathbf{x} = d^2, \tag{10.16}$$

which is an equation of an ellipsoid centered at the origin. If $\mathbf{S}$ is diagonal with unequal diagonal elements, then (10.16) is an equation of a k-dimensional ellipsoid in the standard position. When $k = 2$, ellipsoids become ellipses and spheres become circles.

As we mentioned earlier, an ellipse (a two-dimensional ellipsoid) has two principal axes. By analogy, a k-dimensional ellipsoid has k principal axes; their orientations and lengths are determined by $\mathbf{S}$ and d. The first *principal axis* (also known as the *major axis*) is the straight line that passes through the ellipsoid's greatest dimension. The second principal axis is the straight line that passes through its second greatest dimension, and so on. The last principal axis (also known as the *minor* axis) is the straight line that passes through its shortest dimension. All the axes are pairwise perpendicular to each other.

The manner by which $\mathbf{S}$ and d determine the ellipsoid is given by the following result.

---

2 A symmetric matrix with positive eigenvalues is called a *positive definite* matrix. This and other classifications of matrices are defined in Section 11.3.

Result 10.2. Let **x**, **c**, and **S** be as defined in (10.13), let $\lambda_1 \geq \lambda_2 \geq \ldots \geq \lambda_k > 0$ be the eigenvalues of **S**, and let $\mathbf{v}_1, \mathbf{v}_2, \ldots, \mathbf{v}_k$ be the corresponding normalized eigenvectors. Then the ellipsoid specified by (10.14) has the following characteristics:

(a) The center of the ellipsoid is **c**.
(b) The length of the jth principal axis is $2d\sqrt{\lambda_j}$.
(c) If the jth eigenvalue is of multiplicity 1, then the direction of the jth principal axis is given by $\mathbf{v}_j$ (the ith element of $\mathbf{v}_j$ is the direction cosine of the angle between the jth principal axis and the ith coordinate axis; see Section 4.5.3).
(d) If an eigenvalue is of multiplicity greater than 1, then, by Result 9.3(b), the corresponding eigenvectors are not uniquely defined and, hence, the directions of the corresponding axes are not uniquely defined.
(e) The volume of the ellipsoid is proportional to $d^k \times \det(\mathbf{S})$.

To illustrate, let us apply Result 10.2 to the ellipses drawn in Figures 10.1–10.4. The reader can verify that values defining the circle in Figure 10.1 are

$$\mathbf{c} = \begin{pmatrix} 0 \\ 0 \end{pmatrix}, \qquad \mathbf{S} = \begin{pmatrix} 25 & 0 \\ 0 & 25 \end{pmatrix}, \qquad \text{and} \qquad d = 1. \tag{10.17}$$

In this case, the eigenvalues of **S** are $\lambda_1 = \lambda_2 = 25$. We have an eigenvalue of multiplicity 2. Therefore Result 10.2(d) applies here, and the corresponding eigenvectors are not uniquely defined. In fact, any two orthonormal[3] vectors can be taken as a set of eigenvectors of **S**. For example, the reader can verify that the following sets of vectors can serve as eigenvectors of **S**:

$$\begin{pmatrix} 1 & 0 \\ 0 & 1 \end{pmatrix}, \quad \begin{pmatrix} 0 & 1 \\ 1 & 0 \end{pmatrix}, \quad \begin{pmatrix} 1/\sqrt{2} & 1/\sqrt{2} \\ 1/\sqrt{2} & -1/\sqrt{2} \end{pmatrix}, \quad \begin{pmatrix} 1/\sqrt{5} & 2/\sqrt{5} \\ 2/\sqrt{5} & -1/\sqrt{5} \end{pmatrix}.$$

Confirm that the two vectors in each set are orthonormal and then show that each vector satisfies (9.1).

Notwithstanding the fact that the eigenvectors of **S** in (10.17) are not uniquely defined, we know from Result 10.2(b) that the two axes have the same length of $2d\sqrt{\lambda_1} = 10$, which is the diameter of the circle in Figure 10.1. Unlike the axes of an ellipse, the axes of a circle are not unique. Any straight line that passes through the center of the circle can be considered one axis, and the other axis is taken to be perpendicular to the first.

For the ellipse in Figure 10.2, the eigenvalues and eigenvectors of **S** given in (10.8) are $\lambda_1 = 9$, $\lambda_2 = 4$,

$$\mathbf{v}_1 = \begin{pmatrix} 1 \\ 0 \end{pmatrix}, \quad \text{and} \quad \mathbf{v}_2 = \begin{pmatrix} 0 \\ 1 \end{pmatrix}.$$

---

[3] See Chapter 2 for definition.

By Result 10.2, the length of the major axis is $2d\sqrt{\lambda_1} = 24$ and the length of the minor axis is $2d\sqrt{\lambda_2} = 16$. The major axis is parallel to the horizontal coordinate axis, and the minor axis is parallel to the vertical one. This follows from Result 4.6 because, in this case, the cosine of the angle between $\mathbf{v}_1$ and the horizontal axis (given by the first element of $\mathbf{v}_1$) is 1, which implies that the major axis makes an angle of 0° with the horizontal axis. Also, the cosine of the angle between $\mathbf{v}_1$ and the vertical axis (given by the second element of $\mathbf{v}_1$) is 0, which implies that the major axis makes an angle of 90° with the vertical axis.[4] Similarly, the cosine of the angle between $\mathbf{v}_2$ and the horizontal axis is 0, which implies an angle of 90°, and the cosine of the angle between $\mathbf{v}_2$ and the vertical axis is 1, which implies an angle of 0°.

For the ellipse in Figure 10.3, the eigenvalues and eigenvectors of $\mathbf{S}$ given in (10.11) are $\lambda_1 = 8$, $\lambda_2 = 4$,

$$\mathbf{v}_1 = \frac{1}{2}\begin{pmatrix} \sqrt{3} \\ 1 \end{pmatrix}, \quad \text{and} \quad \mathbf{v}_2 = \frac{1}{2}\begin{pmatrix} -1 \\ \sqrt{3} \end{pmatrix}. \tag{10.18}$$

Thus, by Result 10.2, the lengths of the axes are

$$2d\sqrt{\lambda_1} = 2(3)\sqrt{8} = 16.97 \quad \text{and} \quad 2d\sqrt{\lambda_2} = 2(3)\sqrt{4} = 12,$$

respectively. By Result 4.6, the cosine of the angle between $\mathbf{v}_1$ and the horizontal axis (given by the first element of $\mathbf{v}_1$) is $\sqrt{3}/2$, which implies that the major axis makes an angle of 30° with the horizontal axis. The cosine of the angle between $\mathbf{v}_1$ and the vertical axis (given by the second element of $\mathbf{v}_1$) is 1/2, which implies that the major axis makes an angle of 60° with the vertical axis. Similarly, the cosine of the angle between $\mathbf{v}_2$ and the horizontal axis is $-1/2$, which implies an angle of 120°. The cosine of the angle between $\mathbf{v}_2$ and the vertical axis is $\sqrt{3}/2$, which implies an angle of 30°. Note that $\mathbf{v}_1$ and $\mathbf{v}_2$ are orthogonal, which means that the major and minor axes of the ellipse are perpendicular, as would be expected.

Finally, for the ellipse in Figure 10.4, the eigenvalues and eigenvectors of $\mathbf{S}$ given in (10.12) are $\lambda_1 = 8$, $\lambda_2 = 4$,

$$\mathbf{v}_1 = \frac{1}{2}\begin{pmatrix} 1 \\ \sqrt{3} \end{pmatrix}, \quad \text{and} \quad \mathbf{v}_2 = \frac{1}{2}\begin{pmatrix} \sqrt{3} \\ -1 \end{pmatrix}. \tag{10.19}$$

The eigenvalues of the matrices in (10.11) and (10.12) are the same, indicating that the lengths of the corresponding axes of the ellipses in Figures 10.3 and 10.4 are equal. Notice, however, that the elements in each vector are switched, which indicates that the axes of the ellipses are also switched (the major axis of one has the same direction as the minor axis of the other).

Equations of ellipsoids appear frequently in many applications. For example, in statistics we frequently encounter ellipsoids such as

$$(\boldsymbol{\mu} - \bar{\mathbf{x}})^{\mathrm{T}}\mathbf{S}^{-1}(\boldsymbol{\mu} - \bar{\mathbf{x}}) = d^2, \tag{10.20}$$

---

4 Table 4.1 displays the cosines of some angles.

where $\bar{\mathbf{x}}$ and $\mathbf{S}$ are the mean vector and variance-covariance matrix of a data matrix $\mathbf{X}$, as defined in (8.19). For an appropriate value of d, (10.20) represents the boundary of a confidence ellipsoid for μ, the mean vector of a multivariate population (see any multivariate analysis book, e.g., Johnson and Wichern (1992) or Rencher (1995)).

Another example of an ellipsoid, which is encountered in regression analysis (see Section 8.3), is given by

$$(\beta - \hat{\beta})^{\mathrm{T}}\mathbf{S}^{-1}(\beta - \hat{\beta}) = d^2, \tag{10.21}$$

where $\hat{\beta}$ is the vector of estimated regression coefficients, as defined in (8.39), and $\mathbf{S}$ is the estimated variance-covariance matrix of $\hat{\beta}$. For an appropriate value of d, (10.21) represents the boundary of a confidence ellipsoid for the vector of regression parameters β (see, for example, any one of the regression analysis books listed at the end of Section 8.3).

We shall also encounter ellipsoidal equations in the next two sections when we measure the distance between two points in a space relative to the scatter of other points in the space and when we find the minimum-volume ellipsoid containing a specified number of points in a space.

10.2. Distances

As we mentioned in Section 4.5.2, the concept of distance plays an important role in many statistical applications such as multivariate and regression analysis. In this section we discuss ways of measuring the distance between points in a space. Consider, for example, the data set

$$\mathbf{X} = \begin{pmatrix} 178 & 76 \\ 180 & 80 \\ 170 & 70 \\ 172 & 74 \\ 175 & 75 \end{pmatrix}, \tag{10.22}$$

where the columns represent heights (in centimeters) and weights (in kilograms), respectively, of five individuals as reported in (8.14). In Section 8.2, we computed some summary statistics for this data set. Our objective now is to compare the five individuals with each other. In particular, we wish to measure the magnitude of the difference between any two of the five individuals. Let $\mathbf{x}_i^{\mathrm{T}}$ be the ith row of $\mathbf{X}$. Thus, each vector $\mathbf{x}_i$ consists of two elements, the height and the weight of the ith individual. We are interested in measuring the differences between $\mathbf{x}_i$ and $\mathbf{x}_j$ for all $i \neq j$.

We have seen in Section 4.2 that any algebraic vector can be represented as a point in a space. Since each row of $\mathbf{X}$ consists of two elements, we need a two-dimensional space to represent this data set. Representing each individual by a point in this space, we construct a two-dimensional scatter plot of the variable weight versus the variable height. This scatter plot is shown in Figure 10.6. Please ignore the superimposed circle for the moment.

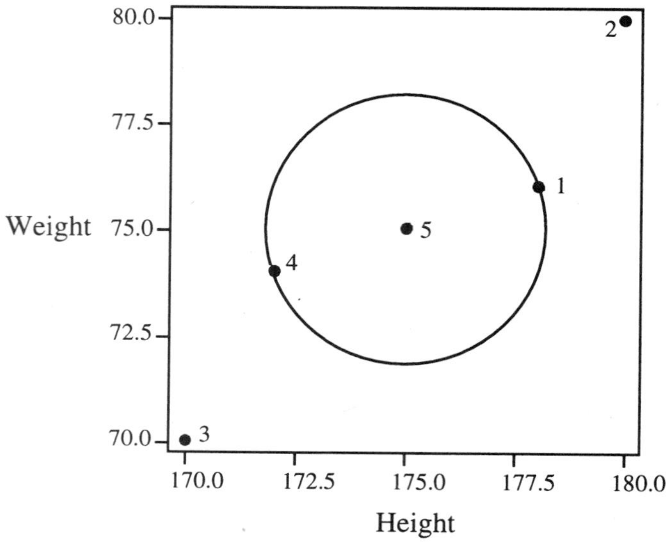

Figure 10.6. Scatter plot of weight versus height for the data matrix **X** given in (10.22) with a superimposed circle of radius 3.16 centered at the sample mean $\bar{\mathbf{x}}$.

The representation of algebraic vectors as points in a space facilitates measuring the magnitude of the difference between two vectors. This magnitude may be measured by the distance between the two points representing the vectors in the space. However, there are many ways of measuring the distance between two points in a space. In this section we define two such ways, namely, the *Euclidean distance* and the *elliptical distance.*

10.2.1. The Euclidean Distance

As we mentioned in Section 4.5.2, a natural way of measuring the distance between two vectors is to take the norm of their difference. The norm of the difference between two vectors is the same as the *Euclidean* or *straight-line distance* between the two points in the space. This general fact is stated in the following result.

Result 10.3. Let **a** and **b** be any vectors of dimensions $k \times 1$. The *Euclidean distance* between **a** and **b** is the norm of the difference between the two vectors, namely,

$$\|\mathbf{a} - \mathbf{b}\| = \sqrt{|a_1 - b_1|^2 + |a_2 - b_2|^2 + \cdots + |a_k - b_k|^2} \qquad (10.23)$$

$$= \sqrt{(\mathbf{a}-\mathbf{b})^{\mathrm{T}}(\mathbf{a}-\mathbf{b})}\,. \tag{10.24}$$

A corollary of Result 10.3 is obtained when $\mathbf{b} = \mathbf{0}$. In this case (10.24) becomes

$$\|\mathbf{a}-\mathbf{0}\| = \sqrt{\mathbf{a}^{\mathrm{T}}\mathbf{a}} = \|\mathbf{a}\|, \tag{10.25}$$

hence the norm of a vector can be interpreted as the distance between the vector (as a point in a space) and the origin (see Section 4.3).

Using Result 10.3, the Euclidean distance between the ith and jth rows of $\mathbf{X}$ is

$$\|\mathbf{x}_i-\mathbf{x}_j\| = \sqrt{(\mathbf{x}_i-\mathbf{x}_j)^{\mathrm{T}}(\mathbf{x}_i-\mathbf{x}_j)}\,. \tag{10.26}$$

For example, the difference between the first and second individuals in the data matrix $\mathbf{X}$ of (10.22) is

$$\mathbf{x}_1-\mathbf{x}_2 = \begin{pmatrix}178\\76\end{pmatrix} - \begin{pmatrix}180\\80\end{pmatrix} = \begin{pmatrix}-2\\-4\end{pmatrix},$$

and the magnitude of this difference can be measured by

$$\|\mathbf{x}_1-\mathbf{x}_2\| = \sqrt{(\mathbf{x}_1-\mathbf{x}_2)^{\mathrm{T}}(\mathbf{x}_1-\mathbf{x}_2)} = \sqrt{(-2\quad -4)\begin{pmatrix}-2\\-4\end{pmatrix}} = \sqrt{20} \cong 4.5.$$

It should be clear that $\|\mathbf{x}_2-\mathbf{x}_1\| = \|\mathbf{x}_1-\mathbf{x}_2\|$; that is, the order in which we take the difference is immaterial. Similarly, we compute the Euclidean distances for each other pair of individuals in the sample. These distances are computed and organized in the matrix:

$$\mathbf{D} = \begin{pmatrix} 0.0 & 4.5 & 10.0 & 6.3 & 3.2\\ 4.5 & 0.0 & 14.1 & 10.0 & 7.1\\ 10.0 & 14.1 & 0.0 & 4.5 & 7.1\\ 6.3 & 10.0 & 4.5 & 0.0 & 3.2\\ 3.2 & 7.1 & 7.1 & 3.2 & 0.0 \end{pmatrix}. \tag{10.27}$$

Thus, $d_{ij} = \|\mathbf{x}_i-\mathbf{x}_j\|$ is the Euclidean distance between the ith and jth individuals. Note that $\mathbf{D}$ is necessarily symmetric because $d_{ij} = d_{ji}$. Also, note that $d_{ii} = 0$, indicating that the distance between an individual and himself is 0. In applications such as cluster analysis,[5] $\mathbf{D}$ is referred to as the *dissimilarity* matrix because its elements measure how dissimilar two individuals are.

---

5 Cluster analysis is used to classify n objects into k groups in such a way that members in the same group are as similar to each other as possible, whereas members of different groups are as dissimilar as possible. See any book on cluster analysis; for example, Hartigan (1975), Romesburg (1984), or Kaufman and Rousseeuw (1990).

In some applications, we are interested in finding the distance between an individual in the data set and a fixed point (usually a typical or average individual). For example, what is the distance between each row of $\mathbf{X}$ and the mean $\bar{\mathbf{x}} = n^{-1}\mathbf{X}^{\mathrm{T}}\mathbf{1}$ as defined in (8.21)? Setting $\mathbf{a} = \mathbf{x}_i$ and $\mathbf{b} = \bar{\mathbf{x}}$, (10.24) becomes

$$\|\mathbf{x}_i - \bar{\mathbf{x}}\| = \sqrt{(\mathbf{x}_i - \bar{\mathbf{x}})^{\mathrm{T}}(\mathbf{x}_i - \bar{\mathbf{x}})}\,. \tag{10.28}$$

A large value of $\|\mathbf{x}_i - \bar{\mathbf{x}}\|$ indicates that $\mathbf{x}_i$ is far from the mean in the Euclidean distance sense. Using (10.28), the reader can verify that the distances between each row of $\mathbf{X}$ in (10.22) and $\bar{\mathbf{x}} = (175 \quad 75)^{\mathrm{T}}$ are

$$(3.16 \quad 7.07 \quad 7.07 \quad 3.16 \quad 0). \tag{10.29}$$

Thus, the second and third individuals are farthest from the sample mean and the fifth individual is closest to the sample mean. The scatter plot of weight versus height in Figure 10.6 shows a superimposed circle centered at $\bar{\mathbf{x}}$ with radius $r = 3.16$. As you can see, the first and fourth points lie exactly on the circumference of the circle because $\|\mathbf{x}_1 - \bar{\mathbf{x}}\| = \|\mathbf{x}_4 - \bar{\mathbf{x}}\| = r$, whereas the second and third points are located outside the circle because both $\|\mathbf{x}_2 - \bar{\mathbf{x}}\|$ and $\|\mathbf{x}_3 - \bar{\mathbf{x}}\|$ are larger than r. The fifth point has a Euclidean distance from $\bar{\mathbf{x}}$ equal to 0, indicating that $\mathbf{x}_5$ must be identical to $\bar{\mathbf{x}}$. An examination of the matrix $\mathbf{X}$ shows that the fifth row is (175 75), which is indeed equal to $\bar{\mathbf{x}}$.

We end this section by noting that (10.23) is a special case of a more general form of a distance, which is given by

$$\|\mathbf{a} - \mathbf{b}\|_p = \left\{|a_1 - b_1|^p + |a_2 - b_2|^p + \cdots + |a_k - b_k|^p\right\}^{1/p}, \tag{10.30}$$

where $p \geq 1$. This distance, known as the *Minkowski* distance, is simply the p-norm of the difference between the vectors $\mathbf{a}$ and $\mathbf{b}$.[6] When $p = 2$, (10.30) reduces to the Euclidean distance in (10.23), and when $p = 1$, the Minkowski distance is known as the *Manhattan* distance. Other types of distances can, of course, be defined. Examples are given in the next section.

10.2.2. The Elliptical Distance

Although the Euclidean distance is a natural measure of distance, it may not be appropriate in some applications. For example, consider again the data matrix $\mathbf{X}$ in (10.22) and suppose we measured the heights in meters rather than centimeters. The data matrix becomes

$$\mathbf{Y} = \begin{pmatrix} 1.78 & 76 \\ 1.80 & 80 \\ 1.70 & 70 \\ 1.72 & 74 \\ 1.75 & 75 \end{pmatrix}.$$

6 See Section 5.3 for a discussion of the p-norm.

The reader can verify that the Euclidean distances for the data set **Y** are

$$\mathbf{W} = \begin{pmatrix} 0 & 4 & 6 & 2 & 1 \\ 4 & 0 & 10 & 6 & 5 \\ 6 & 10 & 0 & 4 & 5 \\ 2 & 6 & 4 & 0 & 1 \\ 1 & 5 & 5 & 1 & 0 \end{pmatrix}. \tag{10.31}$$

Although the matrices **X** and **Y** represent the same measurements, the distances **D** and **W** in (10.27) and (10.31) are not equal either absolutely or relatively. For example, $d_{14} > d_{34}$, but $w_{14} < w_{34}$. This is, of course, troublesome.

The reason the Euclidean distance gives disappointing results in this case is that it gives equal weight to each coordinate, which causes it to be sensitive to changes in the scale of measurements. One way out of this problem is to first scale the columns of **X** before computing the Euclidean distances. The scaling can be done by dividing each element in a column by the standard deviation of the column.[7]

In this way, the scaled matrix will always be the same no matter which scale we use to measure the data. The variances of the first and second columns of **X** are found in (8.18) to be 17 and 13, respectively. Dividing each column by its standard deviation yields the scaled version of **X**, namely,

$$\mathbf{Z} = \begin{pmatrix} 43.17 & 21.08 \\ 43.66 & 22.19 \\ 41.23 & 19.41 \\ 41.72 & 20.52 \\ 42.44 & 20.80 \end{pmatrix}. \tag{10.32}$$

The scaled version of **Y** is the same as that of **X**. The fact that the two scaled versions are equal is expected because scaling gets rid of the effects of changes in the scale of measurements. Again, the distance between the ith and jth rows of **Z** is

$$\|\mathbf{z}_i - \mathbf{z}_j\| = \sqrt{(\mathbf{z}_i - \mathbf{z}_j)^{\mathrm{T}}(\mathbf{z}_i - \mathbf{z}_j)}. \tag{10.33}$$

Using (10.33), we compute the Euclidean distances between the rows of **Z** and organize them in the matrix:

$$\mathbf{D} = \begin{pmatrix} 0.0 & 1.2 & 2.6 & 1.6 & 0.8 \\ 1.2 & 0.0 & 3.7 & 2.6 & 1.8 \\ 2.6 & 3.7 & 0.0 & 1.2 & 1.8 \\ 1.6 & 2.6 & 1.2 & 0.0 & 0.8 \\ 0.8 & 1.8 & 1.8 & 0.8 & 0.0 \end{pmatrix}. \tag{10.34}$$

---

7 Recall that the standard deviation is the positive square root of the variance as defined in (8.17).

The matrix $\mathbf{Z}$ is obtained from either $\mathbf{X}$ or $\mathbf{Y}$ by dividing each column by its standard deviation before computing the Euclidean distances. From Chapter 2, we know that post-multiplying a matrix by a diagonal matrix is the same as multiplying its columns by the corresponding diagonal elements. Therefore, the standardized version $\mathbf{Z}$ of any matrix $\mathbf{X}$ can be written as

$$\mathbf{Z} = \mathbf{X}\mathbf{A}^{-1},$$

where $\mathbf{A} = \text{diag}(a_{jj})$ is a diagonal matrix and a_{jj} is the standard deviation of the jth column of $\mathbf{X}$. Thus, the ith row of $\mathbf{Z}$ is $\mathbf{z}_i^{\mathrm{T}} = \mathbf{x}_i^{\mathrm{T}}\mathbf{A}^{-1}$ and hence $\mathbf{z}_i = \mathbf{A}^{-1}\mathbf{x}_i$. Similarly, $\mathbf{z}_j = \mathbf{A}^{-1}\mathbf{x}_j$. Substituting $\mathbf{z}_i$ and $\mathbf{z}_j$ in (10.26), we obtain

$$\begin{aligned}
\|\mathbf{z}_i - \mathbf{z}_j\| &= \sqrt{(\mathbf{z}_i - \mathbf{z}_j)^{\mathrm{T}}(\mathbf{z}_i - \mathbf{z}_j)} \\
&= \sqrt{\left(\mathbf{A}^{-1}\mathbf{x}_i - \mathbf{A}^{-1}\mathbf{x}_j\right)^{\mathrm{T}}\left(\mathbf{A}^{-1}\mathbf{x}_i - \mathbf{A}^{-1}\mathbf{x}_j\right)} \\
&= \sqrt{\left(\mathbf{A}^{-1}(\mathbf{x}_i - \mathbf{x}_j)\right)^{\mathrm{T}}\left(\mathbf{A}^{-1}(\mathbf{x}_i - \mathbf{x}_j)\right)} \\
&= \sqrt{(\mathbf{x}_i - \mathbf{x}_j)^{\mathrm{T}}\mathbf{B}^{-1}(\mathbf{x}_i - \mathbf{x}_j)}, \qquad (10.35)
\end{aligned}$$

where $\mathbf{B} = \mathbf{A}^2$. The matrix $\mathbf{B}$ is a diagonal matrix whose jth diagonal element is the squared standard deviation or, simply, the variance of the jth column of $\mathbf{X}$; that is, $b_{jj} = s_{jj}$, where s_{jj} is as defined in (8.17).

The advantage of (10.35) over (10.26) is that the former is insensitive to the scale of measurements. The distance in (10.35), however, still needs one further modification. This measure is appropriate if the columns of $\mathbf{X}$ are not related, but in our example, one would expect that an individual's weight and height are in fact related. Therefore we have to take into account the relationship between the columns of $\mathbf{X}$ when we compute the distances. Recall from Section 8.2.2 that the relationships among the columns of $\mathbf{X}$ are summarized in the variance-covariance matrix $\mathbf{S}$ defined in (8.19). Thus, to account for such relationships, we replace $\mathbf{B}$ in (10.35) by the variance-covariance matrix $\mathbf{S}$. This yields

$$d(\mathbf{x}_i, \mathbf{x}_j) = \sqrt{(\mathbf{x}_i - \mathbf{x}_j)^{\mathrm{T}}\mathbf{S}^{-1}(\mathbf{x}_i - \mathbf{x}_j)}, \qquad (10.36)$$

provided that $\mathbf{S}$ is nonsingular. If $\mathbf{S}$ is singular, it is still possible to compute a distance similar to the one defined for nonsingular matrices in (10.36); see Hadi (1992) for the details.

We refer to the distance in (10.36) as the *elliptical* distance because if we take the square of both sides of (10.36), we obtain an equation of an ellipsoid as in (10.14). The *elliptical* distance is sometimes referred to as the *relative* distance because it is measured relative to the variance-covariance structure of the matrix $\mathbf{X}$. In other words, the distance between two points in the space is measured relative to the scatter of all points in the space. Note that, to distinguish between (10.35) and (10.36), the latter is denoted by $d(\mathbf{x}_i, \mathbf{x}_j)$.

This type of distance is the one most commonly encountered in statistics and data analysis. For example, when $\mathbf{x}_j$ is replaced by the sample mean $\overline{\mathbf{x}}$, (10.36) becomes

$$d(\mathbf{x}_i, \overline{\mathbf{x}}) = \sqrt{(\mathbf{x}_i - \overline{\mathbf{x}})^{\mathrm{T}}\mathbf{S}^{-1}(\mathbf{x}_i - \overline{\mathbf{x}})}, \tag{10.37}$$

which is known in the statistical literature as the *Mahalanobis* distance. It measures the distance between $\mathbf{x}_i$ and $\overline{\mathbf{x}}$ relative to the variance-covariance matrix $\mathbf{S}$. Thus, according to the Mahalanobis distance, all points lying on the surface of the ellipsoid specified by (10.37) are considered to be equally distant from $\overline{\mathbf{x}}$.

Using (10.37), the Mahalanobis distances computed from the matrix $\mathbf{X}$ in (10.22) are found to be

$$(1.41 \quad 1.41 \quad 1.41 \quad 1.41 \quad 0). \tag{10.38}$$

Thus, the first four individuals lie on an ellipse specified by the equation

$$(\mathbf{x}_i - \overline{\mathbf{x}})^{\mathrm{T}}\mathbf{S}^{-1}(\mathbf{x}_i - \overline{\mathbf{x}}) = (1.41)^2. \tag{10.39}$$

This ellipse is superimposed on the scatter plot of weight versus height in Figure 10.7. A circle centered at $\overline{\mathbf{x}}$ with a radius of 1.41 is also shown. Recall from Result 10.2 that the directions of the major and minor axes of this ellipse are determined by the first and second eigenvectors of $\mathbf{S}$, and that the lengths of the major and minor axes are $2(1.41)\sqrt{\lambda_1}$ and $2(1.41)\sqrt{\lambda_2}$, respectively.

According to (10.28), points at the circumference of the circle are considered to be equally distant from $\overline{\mathbf{x}}$, whereas according to (10.39), points at the circumference of the ellipse are considered to be equally distant from $\overline{\mathbf{x}}$. It might appear at first that the distances in (10.38) are counterintuitive; points 1 and 4, for example, are physically closer to $\overline{\mathbf{x}}$ than points 2 and 3. But the elliptical distance takes into account the relationship between weight and height as well as their variability. Because the two variables are measured using different units, weight turns out to be less variable than height. Also, the data indicate that there is a positive correlation ($r = 0.94$) between the two variables. Therefore, the elliptical distance makes perfect sense in situations such as the present one.

The Mahalanobis distance was originally proposed for the detection of outliers in multivariate data, but it sometimes fails as an outlier detector. For other methods for detecting multivariate outliers, see, e.g., Rousseeuw and van Zomeren (1990); Hadi (1992, 1994); Gould and Hadi (1993); Hadi and Simonoff (1993); and Peña and Yohai (1994).

In other applications, it may be appropriate to substitute some other matrix $\mathbf{A}$ for $\mathbf{S}$, in which case the distances would be computed relative to $\mathbf{A}$. Accordingly, the elliptical distance can be defined in its general form as given in the next result.

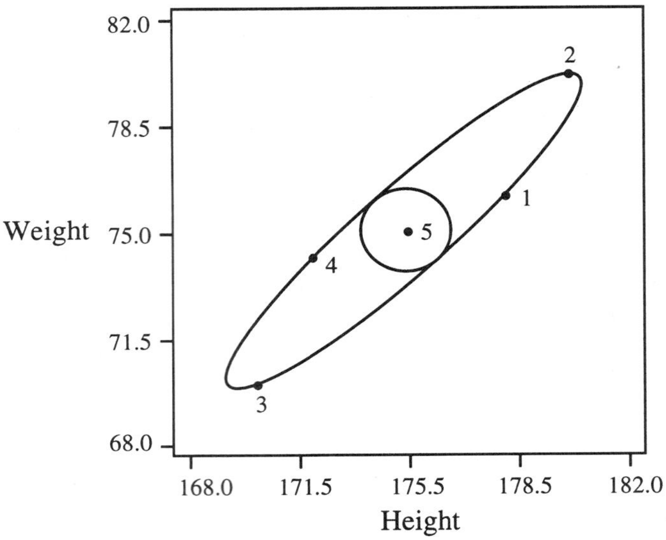

Figure 10.7. Scatter plot of weight versus height as in Figure 10.6. The superimposed circle and ellipse are both centered at the sample mean $\bar{\mathbf{x}}$. The circle has a radius of 1.41, and the ellipse is determined by the variance-covariance matrix **S** and the constant $d^2 = (1.41)^2$.

Result 10.4. The elliptical distance between any two vectors **a** and **b** relative to a matrix **A** is given by

$$d(\mathbf{a}, \mathbf{b}) = \sqrt{(\mathbf{a} - \mathbf{b})^T \mathbf{A}^{-1} (\mathbf{a} - \mathbf{b})}, \tag{10.40}$$

as long as **A** is symmetric with positive eigenvalues.
Note that when $\mathbf{b} = \mathbf{0}$, (10.40) becomes

$$d(\mathbf{a}, \mathbf{0}) = \sqrt{\mathbf{a}^T \mathbf{A}^{-1} \mathbf{a}}. \tag{10.41}$$

The elliptical distance in (10.41) is sometimes referred to as the *semi-norm* because when $\mathbf{A} = \mathbf{I}$, (10.41) reduces to (10.25), which is the norm of **a**. Several other special cases of (10.40) can be obtained. For example, when $\mathbf{A} = \mathbf{I}$, (10.40) reduces to (10.24); when $\mathbf{A} = \mathbf{S}$, it reduces to (10.36); and when **S** is diagonal, it reduces to (10.35).

In principle, (10.40) can be computed for any nonsingular matrix **A**. However, we require **A** to be symmetric and to have positive eigenvalues so that the distance defined in (10.40) will have some nice properties, as stated in the following result.

Result 10.5. Let $d(\mathbf{a}, \mathbf{b})$ be the elliptical distance defined in (10.40). Then $d(\mathbf{a}, \mathbf{b})$ satisfies the following intuitively desirable properties:

(a) $d(\mathbf{a}, \mathbf{b}) \geq 0$ for all $\mathbf{a}$ and $\mathbf{b}$ (a distance cannot be negative).

(b) $d(\mathbf{a}, \mathbf{b}) = 0$ if and only if $\mathbf{a} = \mathbf{b}$ (the distance between a vector and itself is 0).

(c) $d(\mathbf{a}, \mathbf{b}) = d(\mathbf{b}, \mathbf{a})$ for all $\mathbf{a}$ and $\mathbf{b}$ (the distance treats the two vectors symmetrically).

(d) $d(\mathbf{a}, \mathbf{b}) \leq d(\mathbf{a}, \mathbf{c}) + d(\mathbf{b}, \mathbf{c})$ for all $\mathbf{a}$, $\mathbf{b}$, and $\mathbf{c}$ (the triangle inequality; similar to Result 5.5(d)).

Note that all of the distances discussed in this section satisfy the above properties because they are special cases of (10.40).

*10.3. Smallest Ellipsoid Containing a Scatter of Points

We have seen in (10.14) that a k-dimensional ellipsoid is determined by three quantities: the vector $\mathbf{c}$ determines its center, the matrix $\mathbf{S}$ determines its orientation (the directions of its principal axes), and the scalar d^2, together with $\mathbf{S}$, determines its orientation and shape and hence its size. The size of an ellipse ($k = 2$) is measured by its area, and the size of an ellipsoid ($k > 2$) is measured by its volume. In this section, we answer the question: Given $\mathbf{c}$ and $\mathbf{S}$, and an $n \times k$ matrix $\mathbf{X}$ whose rows can be plotted as points in a k-dimensional space, what is the smallest ellipsoid determined by $\mathbf{c}$ and $\mathbf{S}$ that contains all or a specified number of the n points? As we discussed in Section 10.1, for fixed $\mathbf{S}$, the volume of an ellipsoid is completely determined by d^2; the larger d^2, the larger the volume. Thus, d^2 is the key to answering the above question.

Let us first start with the simple case where $k = 2$. One strategy for finding the smallest ellipse that contains all the n points is to first plot the data, then draw an ellipse (with center and orientation determined by $\mathbf{c}$ and $\mathbf{S}$) for an initial value of d^2. We have three possible situations:

(a) The ellipse contains n points, and at least one point is on its circumference. In this case, this is clearly the desired ellipse.

(b) The ellipse contains n points but no point is lying on its circumference. In this case, we decrease the value of d^2 (i.e., deflate the ellipse) until the ellipse satisfies condition (a).

(c) The ellipse contains fewer than n points. In this case, we increase the value of d^2 (i.e., inflate the ellipse) until the ellipse satisfies condition (a).

Similarly, to find the smallest ellipse that contains at least $p < n$ points, we follow the same strategy but with n replaced by p.

Actually, we can implement the above strategy numerically (without having to draw ellipses) and generalize it to situations where $k > 2$ as follows. For each point $\mathbf{x}_i$, we replace $\mathbf{x}$ in (10.14) by $\mathbf{x}_i$ and obtain

$$a_i^2 = (\mathbf{x}_i - \mathbf{c})^{\mathrm{T}}\mathbf{S}^{-1}(\mathbf{x}_i - \mathbf{c}), \quad i = 1, 2, \ldots, n. \tag{10.42}$$

Thus, a_i is the distance between $\mathbf{x}_i$ and $\mathbf{c}$ relative to $\mathbf{S}$. Setting

$$d^2 = \max\,(a_1^2, a_2^2, \ldots, a_n^2),$$

then the ellipsoid

$$(\mathbf{x}_i - \mathbf{c})^{\mathrm{T}}\mathbf{S}^{-1}(\mathbf{x}_i - \mathbf{c}) = d^2 \tag{10.43}$$

must contain all the n points, and at least one point (the point whose a_i^2 value is equal to d^2) is on the surface of the ellipsoid. Therefore (10.43) is the smallest ellipsoid containing all n points.

To find the smallest ellipsoid that contains at least $p < n$ points, we proceed in the same way as above. So, we compute (10.42) and let d^2 in (10.43) be the pth smallest value of $a_1^2, a_2^2, \ldots, a_n^2$.

We illustrate the above procedures with two examples. For the first example, we use the matrix $\mathbf{X}$ in (10.22) and let $\mathbf{c} = \bar{\mathbf{x}}$ and let $\mathbf{S}$ be the variance-covariance matrix of $\mathbf{X}$. The a_i^2 in (10.42) are found to be

$$\begin{pmatrix} 2 & 2 & 2 & 2 & 0 \end{pmatrix}. \tag{10.44}$$

By comparing (10.37) to (10.42), we see that the a_i^2 in (10.44) are the squares of the Mahalanobis distances in (10.38). In this case, $d^2 = \max(a_i^2) = 2$. Hence, taking $\bar{\mathbf{x}}$ and $\mathbf{S}$ as given, the smallest ellipse containing all five points is specified by $\bar{\mathbf{x}}$, $\mathbf{S}$, and d^2. This ellipse is drawn in Figure 10.7. Note that there are four points on the circumference of the ellipse.

For the second example, suppose in addition to the five individuals whose weights and heights are given by $\mathbf{X}$ in (10.22) we sampled four more individuals and measured their heights and weights. The heights and weights of all nine individuals are

$$\mathbf{X} = \begin{pmatrix} 178 & 76 \\ 180 & 80 \\ 170 & 70 \\ 172 & 74 \\ 175 & 75 \\ 171 & 73 \\ 179 & 78 \\ 176 & 75 \\ 174 & 74 \end{pmatrix}. \tag{10.45}$$

Now, suppose we wish to find the smallest ellipse that is specified by

$$\mathbf{c} = \begin{pmatrix} 175 \\ 75 \end{pmatrix} \quad \text{and} \quad \mathbf{S} = \begin{pmatrix} 12.750 & 9.625 \\ 9.625 & 8.250 \end{pmatrix} \tag{10.46}$$

and that contains at least four points. Note that while the vector $\mathbf{c}$ in (10.46) coincides with the mean of the columns of $\mathbf{X}$, the matrix $\mathbf{S}$ in (10.46) is not the variance-covariance matrix of $\mathbf{X}$. Here, both $\mathbf{c}$ and $\mathbf{S}$ are given.

So, for each point in $\mathbf{X}$, we need to compute

$$a_i^2 = (\mathbf{x}_i - \mathbf{c})^T \mathbf{S}^{-1} (\mathbf{x}_i - \mathbf{c}), \quad i = 1, 2, \ldots, 9, \tag{10.47}$$

which gives

$$\begin{pmatrix} 2.33 & 3.49 & 3.49 & 2.33 & 0.00 & 2.31 & 1.26 & 0.66 & 0.14 \end{pmatrix}. \tag{10.48}$$

The fourth smallest value is 1.26, corresponding to the seventh row of $\mathbf{X}$. Therefore, the smallest ellipse containing at least four points is specified by

$$\mathbf{c} = \begin{pmatrix} 175 \\ 75 \end{pmatrix}, \quad \mathbf{S} = \begin{pmatrix} 12.750 & 9.625 \\ 9.625 & 8.250 \end{pmatrix}, \quad \text{and} \quad d^2 = 1.26. \tag{10.49}$$

This ellipse (shown in Figure 10.8) contains exactly four points (5, 7, 8, 9), and point 7 (179, 78) lies on the circumference of the ellipse.

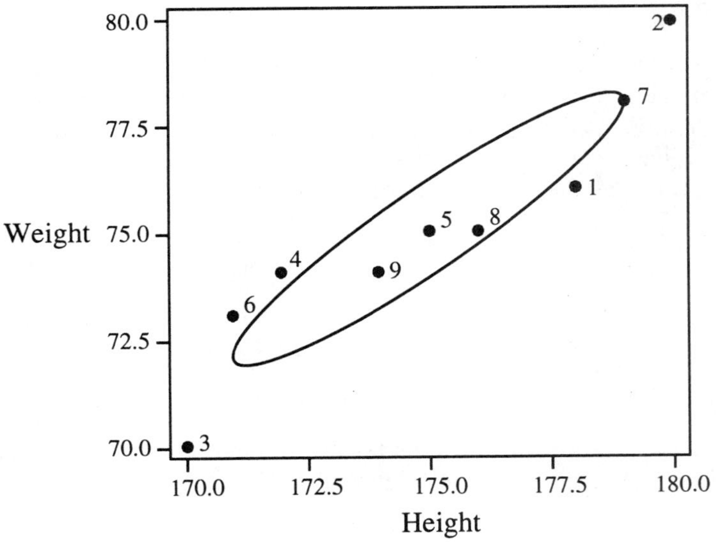

Figure 10.8. Scatter plot of the columns of $\mathbf{X}$ in (10.45) with the smallest ellipse that contains exactly four points. This ellipse is defined by (10.49).

Given $\mathbf{c}$ and $\mathbf{S}$, it is therefore straightforward to find the smallest ellipsoid that contains at least $p \leq n$ points. However, if $\mathbf{c}$ and $\mathbf{S}$ are not given and we wish to find the smallest ellipsoid that contains a specified number of points, the problem becomes much more difficult. For example, suppose we are given an $n \times k$ matrix ($k < n$) representing n points in a k-dimensional space. Let m be a subset of points, of size $p > k$, and let $\bar{\mathbf{x}}_m$ and $\mathbf{S}_m$ be the mean and variance-covariance matrix of the points in subset m. Provided that $\mathbf{S}_m$ is nonsingular, the smallest ellipsoid specified by $\bar{\mathbf{x}}_m$ and $\mathbf{S}_m$ that contains at least p points is given by

$$(\mathbf{x}_i - \bar{\mathbf{x}}_m)^{\mathrm{T}} \mathbf{S}_m^{-1} (\mathbf{x}_i - \bar{\mathbf{x}}_m) = d^2, \tag{10.50}$$

where d^2 is the smallest value for which the ellipsoid contains at least p points. Let V_m be the volume of this ellipsoid. We wish to find the subset m for which V_m is smallest. The ellipsoid based on this subset is known in the statistical literature as the minimum volume ellipsoid (MVE) (see Rousseeuw and Leroy (1987); Hadi and Simonoff (1994)). The MVE has important implications in the theory of statistical estimation and in the detection of outlying or extreme observations, two important theoretical and practical problems.

Now, how do we find the subset with the MVE? Given the n points, one can obtain

$$\frac{n!}{p!(n-p)!} \tag{10.51}$$

possible subsets of size p, where $n!$ (read as n-factorial) is the product of the first n positive integers, that is, $n! = 1 \times 2 \times \cdots \times (n-2) \times (n-1) \times n$. The formula in (10.51) is known as the *combination* formula because it represents the number of possible combinations of p objects that can be constructed from a total of n objects. For example, for $n = 9$ and $p = 4$, as in the above example, there are

$$\frac{n!}{p!(n-p)!} = \frac{9!}{4!(9-4)!}$$

$$= \frac{1 \times 2 \times 3 \times 4 \times 5 \times 6 \times 7 \times 8 \times 9}{(1 \times 2 \times 3 \times 4)(1 \times 2 \times 3 \times 4 \times 5)} = 126$$

possible subsets of four points that can be constructed out of nine points. Therefore, we can find the subset with the MVE as follows. For each of these 126 subsets of four points, we:

1. Compute the corresponding ellipsoid specified by (10.50).
2. Compute its volume using Result 10.2(e).
3. Choose the subset with the smallest volume.

The above procedure is straightforward. However, its main problem stems from the fact that the number of possible subsets given by (10.51) increases dramatically with n. For example, for $n = 50$ and $p = 5$, the number of possible subsets of five points is 2,118,760; for $n = 50$ and $p = 10$, there are more than 10 billion possible subsets of ten points; and for $n = 50$ and $p = 25$, the number of all possible subsets is astronomical. Therefore, for large n the problem becomes practically impossible. Approximate, but practically feasible, solutions to this problem are suggested by Rousseeuw and van Zomeren (1990) and Hadi (1992, 1994).

Exercises

10.1. Let $\mathbf{x} = (x_1 \;\; x_2)^{\mathrm{T}}$. For each of the following expressions, find the matrix $\mathbf{S} = \mathbf{S}^{\mathrm{T}}$ so that the expression can be written as $\mathbf{x}^{\mathrm{T}}\mathbf{S}^{-1}\mathbf{x}$.

(a) $4x_1^2 + 2x_2^2$ (b) $x_1^2 + 4x_2^2$

(c) $2x_1^2 + 5x_2^2 - 2x_1x_2$ (d) $1.5x_1^2 + 1.5x_2^2 - x_1x_2$

10.2. Let $\mathbf{x} = (x_1 \;\; x_2)^{\mathrm{T}}$. Write $(\mathbf{x} - \mathbf{c})^{\mathrm{T}}\mathbf{S}^{-1}(\mathbf{x} - \mathbf{c})$ in terms of x_1 and x_2 in each of the following cases.

(a) $\mathbf{S} = \begin{pmatrix} 2 & 0 \\ 0 & 1 \end{pmatrix}$, $\mathbf{c} = \begin{pmatrix} 0 \\ 0 \end{pmatrix}$. (b) $\mathbf{S}^{-1} = \begin{pmatrix} 1.5 & -0.5 \\ -0.5 & 1.5 \end{pmatrix}$, $\mathbf{c} = \begin{pmatrix} 1 \\ 1 \end{pmatrix}$.

(c) $\mathbf{S} = \begin{pmatrix} 5 & 1 \\ 1 & 2 \end{pmatrix}$, $\mathbf{c} = \begin{pmatrix} 0 \\ 0 \end{pmatrix}$. (d) $\mathbf{S}^{-1} = \begin{pmatrix} 0.75 & 0.25 \\ 0.25 & 0.75 \end{pmatrix}$, $\mathbf{c} = \begin{pmatrix} 1 \\ 1 \end{pmatrix}$.

10.3. Find the matrix $\mathbf{S} = \mathbf{S}^{\mathrm{T}}$ such that $\mathbf{x}^{\mathrm{T}}\mathbf{A}\mathbf{x} = \mathbf{x}^{\mathrm{T}}\mathbf{S}\mathbf{x}$ in each of the following cases.

(a) $\mathbf{A} = \begin{pmatrix} 2 & 1 \\ 2 & 4 \end{pmatrix}$. (b) $\mathbf{A} = \begin{pmatrix} 2 & -1 \\ 3 & 4 \end{pmatrix}$.

(c) $\mathbf{A} = \begin{pmatrix} 5 & 0 & 1 \\ 2 & 2 & 0 \\ 3 & 2 & 3 \end{pmatrix}$. (d) $\mathbf{A} = \begin{pmatrix} 1 & 0 & 1 \\ 2 & 2 & 0 \\ 0 & 0 & 1 \end{pmatrix}$.

10.4. Sketch the graph of the function $(\mathbf{x} - \mathbf{c})^{\mathrm{T}}\mathbf{S}^{-1}(\mathbf{x} - \mathbf{c}) = a$ for each of the following cases.

(a) $\mathbf{S}^{-1} = \begin{pmatrix} 4 & 0 \\ 0 & 4 \end{pmatrix}$, $\mathbf{c} = \begin{pmatrix} 0 \\ 0 \end{pmatrix}$, $a = 4$. (b) $\mathbf{S}^{-1} = \begin{pmatrix} 4 & 0 \\ 0 & 16 \end{pmatrix}$, $\mathbf{c} = \begin{pmatrix} 1 \\ 1 \end{pmatrix}$, $a = 1$.

(c) $\mathbf{S}^{-1} = \begin{pmatrix} 5 & 1 \\ 1 & 2 \end{pmatrix}$, $\mathbf{c} = \begin{pmatrix} 0 \\ 0 \end{pmatrix}$, $a = 4$. (d) $\mathbf{S}^{-1} = \begin{pmatrix} 3 & -1 \\ -1 & 3 \end{pmatrix}$, $\mathbf{c} = \begin{pmatrix} 1 \\ 1 \end{pmatrix}$, $a = 1$.

10.5. Consider the ellipses (circles) defined in Exercise 10.4.

(a) Compute the lengths of the axes for each ellipse.

(b) Compute the cosines of the angles between each of the ellipse's axes and the horizontal axis.
(c) What are the angles between each ellipse's axes and the horizontal axis?
(d) Arrange the four ellipses in an increasing order of volume.

10.6. The following data matrix consists of five observations on two variables:

$$\mathbf{X} = \begin{pmatrix} 2 & 2 \\ 2 & 1 \\ 1 & -1 \\ -1 & 1 \end{pmatrix}.$$

For each of these data sets:
(a) Draw a scatter plot of one variable versus the other.
(b) Compute the Manhattan, the Euclidean, and the Mahalanobis distances for each of the five observations and then order the observations accordingly in each case.
(c) Choose a value for p (other than 1 or 2) and compute the Minkowski distance for each observation and order the five observations accordingly.
(d) Without information about the nature of the two variables, can you determine which of the above orderings of the observations would make more sense to you than the others (why)?

10.7. Given a data matrix $\mathbf{X}$ consisting of n observations on k variables, which distance would you use in each of the following situations?
(a) $\mathbf{X}$ consists of two variables, the first of which is the scores of 30 students in the mid-term examination (worth 30 percent of the grade in the course) and the second of which is their corresponding scores in the final examination (worth 70 percent of the grade). We wish to order the students according to their final grade in the course.
(b) $\mathbf{X}$ consists of five variables representing the grades (a number from 0 to 4) in five courses at the end of one semester for each student in a class. We wish to order the students according to their grade point average.
(c) $\mathbf{X}$ consists of two columns; the first contains the ages of 30 students in a given class and the second contains the ages of 30 students in another class.
(d) In an experiment, two fair coins are tossed 100 times each and for each coin the number of heads is observed. The data are organized in a matrix $\mathbf{X}$ which has two columns (one for each coin) and 100 rows (one for each experiment). We wish to know which experiment, if any, has produced the most unexpected outcome.
(e) A random sample of size n_i individuals is taken from country i, $i = 1, 2, \ldots, 30$. In each sample, the percentage of people who smoke (the first variable) and the percentage of people who have lung cancer (the second variable) are recorded.

(f) A random sample consists of 500 families. Each family consists of a father, a mother, and a number of children. For each family, the age of the father, the age of the mother, the number of children, and the family income are recorded. We wish to determine which of the families in this sample is most unusual.
(g) Given the same data as in (f), we wish to measure, for each pair of families in the sample, how similar or dissimilar they are.
(h) $\mathbf{X}$ contains n observations on three variables and we wish to know which of the observations is farthest from the origin in a three-dimensional scatter plot of the columns of $\mathbf{X}$.

10.8. The following are the Mahalanobis distances for each observation in a data matrix $\mathbf{X}$ consisting of three variables:

2.8	1.7	3.5	2.1	2.0
1.3	2.0	2.4	2.5	3.1
2.6	2.0	1.9	2.3	

Let $\bar{\mathbf{x}}$ and $\mathbf{S}$ be the sample mean and variance-covariance matrix, respectively. If possible, write the equation of an ellipsoid determined by $\bar{\mathbf{x}}$ and $\mathbf{S}$ which:
(a) is as small as possible and contains all of the points.
(b) is larger than the ellipsoid in (a).
(c) contains at least half of the points but fewer than 14 points.
(d) is as small as possible and contains exactly half of the points.
(e) is larger than the ellipsoid in (d), yet still contains exactly seven points.
(f) contains exactly five points.
(g) contains no points.

10.9. Compute the dissimilarity matrix for a data set of interest to you, or use

$$\mathbf{X} = \begin{pmatrix} 1 & 0 \\ 0 & 1 \\ 1 & -1 \\ 0 & 0 \end{pmatrix}.$$

10.10. Select and compute a measure of distance using $\mathbf{X}$ in Exercise 10.9. Then verify that your chosen distance satisfies all four properties in Result 10.5.

11 Additional Applications

In Chapters 8 and 10 we presented several applications of matrix algebra. We showed, for example, how matrix notation and concepts can be used to solve simultaneous systems of linear equations and to simplify the formulation of descriptive statistics and linear regression. In this chapter we shall explore additional examples of the applications of matrix algebra. Specifically, we discuss the following applications of the eigenvalue problem:

- Matrix norms
- Collinearity diagnostics
- Classification of quadratic forms
- Markov chains
- Principal component analysis
- Factor analysis

*11.1. Matrix Norms

In Section 5.3, we introduced the concept of vector norms as a way of measuring the magnitude of a vector. Having discussed the singular values of a matrix (see Section 9.3), we are now ready to introduce the concept of matrix norms. Again, the objective here is to measure the magnitude of a given matrix by a single number. In this section, we define two types of matrix norms: the *Frobenius norm* and the *p*-norm.

11.1.1. The Frobenius Norm

The Frobenius norm of any matrix $\mathbf{A}$ of order $m \times n$ is defined by

$$\|\mathbf{A}\|_F = +\sqrt{\sum_{i=1}^{m} \sum_{j=1}^{n} a_{ij}^2}\,. \tag{11.1}$$

In words, the Frobenius norm of a matrix is the positive square root of the sum of its squared elements. As an example, let us compute the Frobenius norm of the matrix

$$\mathbf{A} = \begin{pmatrix} -1 & 1 \\ 2 & 3 \\ 1 & -3 \end{pmatrix}. \tag{11.2}$$

We have $\|\mathbf{A}\|_F = \sqrt{(-1)^2 + 1^2 + 2^2 + 3^2 + 1^2 + (-3)^2} = \sqrt{25} = 5.$

11.1.2. The Matrix P-Norm

Recall from (5.6) that the p-norm of a vector $\mathbf{v}$ is defined as

$$\|\mathbf{v}\|_p = \{|v_1|^p + |v_2|^p + \cdots + |v_n|^p\}^{1/p}$$

where $p \geq 1$. To define the p-norm of a matrix $\mathbf{A}$, we first observe that $\mathbf{Ax}$ is a vector for any conformable vector $\mathbf{x}$, and hence we can represent its p-norm by $\|\mathbf{Ax}\|_p$. The *p-norm* of a matrix $\mathbf{A}$ can then be defined as

$$\|\mathbf{A}\|_p = \max_{\mathbf{x}\neq\mathbf{0}} \frac{\|\mathbf{Ax}\|_p}{\|\mathbf{x}\|_p}. \tag{11.3}$$

The p-norm of a matrix is easier to interpret if it is written in the equivalent form

$$\|\mathbf{A}\|_p = \text{maximum of } \|\mathbf{Ax}\|_p, \text{ subject to } \|\mathbf{x}\|_p = 1.$$

Thus, according to the above definition, to compute $\|\mathbf{A}\|_p$, one needs to first compute $\|\mathbf{Ax}\|_p$ for all possible vectors $\mathbf{x}$ satisfying $\|\mathbf{x}\|_p = 1$, then choose $\mathbf{x}$ that gives the largest value of $\|\mathbf{Ax}\|_p$. Since there are infinitely many vectors $\mathbf{x}$ satisfying $\|\mathbf{x}\|_p = 1$, this is clearly an impossible task. For some special values of p, however, there are easy ways to compute $\|\mathbf{A}\|_p$. The most common special cases correspond to $p = 1$, 2, and ∞. For these cases, the p-norms of $\mathbf{A}$ are given by

$$\|\mathbf{A}\|_1 = \max_j \sum_{i=1}^{m} |a_{ij}|, \tag{11.4}$$

$$\|\mathbf{A}\|_2 = \max_{\mathbf{x}\neq\mathbf{0}} \sqrt{\frac{\mathbf{x}^T\mathbf{A}^T\mathbf{Ax}}{\mathbf{x}^T\mathbf{x}}} = \sqrt{\lambda_1} = \delta_1, \tag{11.5}$$

$$\|\mathbf{A}\|_\infty = \max_i \sum_{j=1}^{n} |a_{ij}|, \tag{11.6}$$

where λ_1 is the maximum eigenvalue of $\mathbf{A}^T\mathbf{A}$ and δ_1 is the maximum singular value of $\mathbf{A}$. Thus, $\|\mathbf{A}\|_1$ is the maximum absolute column sum and $\|\mathbf{A}\|_\infty$ is the maximum absolute row sum. The fact that $\|\mathbf{A}\|_2 = \delta_1$ follows directly from Results 9.3(h) and 9.8(a) because the maximum singular value of $\mathbf{A}$ is equal to the positive square root of the maximum eigenvalue of $\mathbf{A}^T\mathbf{A}$. It is also equal to the positive square root of the maximum eigenvalue of $\mathbf{AA}^T$. Thus, to compute $\|\mathbf{A}\|_2$, we need to compute the singular values of $\mathbf{A}$ and take the maximum one to be $\|\mathbf{A}\|_2$.

Consider, for example, the matrix $\mathbf{A}$ in (11.2). We compute the p-norms as follows. The sums of the absolute values of the columns of $\mathbf{A}$ are 4 and 7, hence $\|\mathbf{A}\|_1 = 7$. Similarly, the sums of the absolute values of the rows of $\mathbf{A}$ are 2, 5, and 4, hence $\|\mathbf{A}\|_\infty = 5$. A computer program was used to compute the singular values of $\mathbf{A}$, which turn out to be 4.39 and 2.39 and hence $\|\mathbf{A}\|_2 = 4.39$. The reader can verify that

$$\mathbf{A}^{\mathrm{T}}\mathbf{A} = \begin{pmatrix} 6 & 2 \\ 2 & 19 \end{pmatrix}$$

and

$$\mathbf{A}\mathbf{A}^{\mathrm{T}} = \begin{pmatrix} 2 & 1 & -4 \\ 1 & 13 & -7 \\ -4 & -7 & 10 \end{pmatrix}.$$

It can also be verified that the maximum eigenvalue of $\mathbf{A}^{\mathrm{T}}\mathbf{A}$ and the maximum eigenvalue of $\mathbf{A}\mathbf{A}^{\mathrm{T}}$ are equal to 19.3 (the square of the maximum singular value of $\mathbf{A}$).

The matrix norms have the same properties as the vector norms. Therefore, Result 5.5 can be restated here, with the vectors **a** and **b** replaced by matrices **A** and **B**. In addition, we have the following result.

Result 11.1. Provided that **A** and **B** are conformable for multiplication, then

$$\|\mathbf{AB}\| \leq \|\mathbf{A}\| \times \|\mathbf{B}\|.$$

*11.2. Collinearity Diagnostics

We have seen in Section 8.3 that matrix notation and concepts are indispensable tools for formulating the linear regression model. In particular, the least squares estimate of β is the solution to the normal equations $\mathbf{X}^{\mathrm{T}}\mathbf{X}\beta = \mathbf{X}^{\mathrm{T}}\mathbf{y}$ given in (8.38). Recall from Section 8.3 that when $\mathrm{r}(\mathbf{X}) = k$, $\mathrm{r}(\mathbf{X}^{\mathrm{T}}\mathbf{X}) = k$ and, consequently, $(\mathbf{X}^{\mathrm{T}}\mathbf{X})^{-1}$ exists. In this case, the unique least squares estimate of β is given in (8.39). On the other hand, if $\mathbf{X}$ is rank-deficient ($\mathrm{r}(\mathbf{X}) < k$), we can no longer use (8.39) because $(\mathbf{X}^{\mathrm{T}}\mathbf{X})^{-1}$ does not exist. In this case, we no longer have a unique solution, which is clearly a problem. This problem is known as the *exact collinearity* or *exact linear dependency* problem because the columns of $\mathbf{X}$ are collinear (note that when $\mathbf{X}$ is rank-deficient, its columns must be linearly dependent, that is, at least one column of $\mathbf{X}$ can be written as a linear combination of the other columns).

As can be seen, exact collinearity is easily detected: We simply compute $\mathrm{r}(\mathbf{X})$, and if $\mathrm{r}(\mathbf{X}) < k$ then we have an exact collinearity problem. The more difficult problem occurs when the columns of $\mathbf{X}$ are not *exactly* but *nearly* collinear. This problem is referred to as the *approximate collinearity* problem (or the *collinearity* problem, for short).

Why is collinearity a problem? How can we detect or diagnose collinearity? Is collinearity always harmful? What variables are involved in a collinearity? What can be done when collinearity occurs? The answers to these and other related questions perhaps require a book-length discussion. In this section we briefly discuss how one can detect collinearity. For further details, the reader is referred to Belsley (1991), which is devoted entirely to this problem, and the references therein.

Condition Indices. A (near) linear dependency among the columns of any matrix $\mathbf{X}$ is generally recognized by a "small" singular value of $\mathbf{X}$, or equivalently by a "small" eigenvalue of $\mathbf{X}^{\mathrm{T}}\mathbf{X}$ (see Result 9.8). (The word *small* is put in quotation marks because it is theoretically difficult to determine when a value is significantly small.) If a singular value of **X** is 0, then **X** is exactly collinear. In most practical applications, however, linear relationships among the columns of **X** are not exact dependencies, and hence small singular values in practice are expected to be “close” to 0 but not exactly 0.

In practice, to determine whether a singular value is close enough to 0, we compute the so-called *condition indices* of the matrix **X**. The condition indices of **X** are defined as

$$\kappa_j = \frac{\delta_1}{\delta_j}, \quad \text{for } j = 1, 2, \ldots, k, \tag{11.7}$$

where δ_j is the jth ordered singular value of **X**. (Recall that the singular values of **X**, $\delta_1, \delta_2, \ldots, \delta_k$, are arranged in descending order by convention.) Notice that the first condition index, κ_1, is always 1 and that the jth condition index, κ_j, is the jth smallest condition index. The largest condition index,

$$\kappa_k = \frac{\delta_1}{\delta_k}, \tag{11.8}$$

is sometimes known as the *condition number* of **X** and is denoted by κ, for simplicity. Thus, the condition number of a matrix is the ratio of its largest and smallest singular values. By Result 9.8(a), the condition number can also be written in terms of the eigenvalues of $\mathbf{X}^T\mathbf{X}$ as

$$\kappa = +\sqrt{\frac{\lambda_1}{\lambda_k}}. \tag{11.9}$$

Now, since a "small" singular value of **X** indicates near linear dependency among the columns of **X**, the near linear dependency can also be characterized by a “large” condition number. One difficulty in using the singular values or, equivalently, the condition indices stems from the fact that the singular values, and hence the condition indices, are sensitive to column scaling. For example, if one of the columns in **X** represents income measured in dollars, then the condition indices will change if, instead, we measured income in thousands of dollars. For this reason and for the purpose of diagnosing collinearity, it is desirable to scale the columns of **X** so that each column has a length (norm) of 1. This is known as *scaling*. Scaling can be accomplished by dividing each element of $\mathbf{x}_j$, the jth column of **X**, by $\|\mathbf{x}_j\|$, the norm (length) of that column. It is easier to interpret the values of the condition indices when **X** is scaled. In this case, empirical evidence (see, e.g., Belsley, 1991) shows that a value of κ_k less than 30 may indicate a weak collinearity, a value between 30 and 100 would indicate a strong collinearity, and a value greater than 100 indicates an extreme collinearity. If collinearity is present, the matrix is said to be *ill-conditioned*.

Furthermore, the number of large condition indices indicates the number of different sets of collinearities. For example, if there are two large condition indices, then there exist two sets of collinearities, each involving a different subset of the columns of **X**. The variables involved in each collinear set can be determined by examining the eigenvectors of $\mathbf{X}^T\mathbf{X}$ corresponding to the large condition indices. For the details, see Belsley (1991).

To summarize, we offer the following steps for diagnosing collinearities involving the columns of any data matrix **X**:

1. If necessary, scale the columns of **X** to unit length.
2. Compute the condition indices of the scaled **X**.
3. Determine the degree of collinearity using the following empirical rule:
   a. if $\kappa_k < 30$, then there is a weak collinearity;
   b. if κ_k is between 30 and 100, then there is a strong collinearity;
   c. if $\kappa_k > 100$, then there is an extreme collinearity;
   d. if κ_k is infinity ($\delta_k = 0$), then there is exact collinearity.
4. The number of different collinearities and the variables involved in each collinear set can be determined by examining the eigenvectors of $\mathbf{X}^T\mathbf{X}$ corresponding to the large condition indices (see below).

We should point out here that there are many issues related to collinearity and collinearity diagnostics that we did not discuss here due to space limitations.[1] The reader is referred to other sources such as Belsley, Kuh, and Welsch (1980); Hadi and Wells (1990); and Belsley (1991).

We conclude this section with a numerical example of the steps above. Consider the following data matrix:

$$\mathbf{X} = \begin{pmatrix} 0.322 & 0.5 & -0.5 & 0.0 \\ 0.012 & 0.5 & 0.5 & -0.5 \\ 0.631 & 0.5 & 0.0 & 0.5 \\ 0.316 & 0.0 & 0.5 & 0.5 \\ 0.000 & 0.0 & 0.5 & 0.0 \\ 0.631 & 0.5 & 0.0 & 0.5 \end{pmatrix}.$$

The reader can verify that each column of **X** has unit length, so the first step is unnecessary in this case. With the aid of a computer, it is found that the singular values of **X** are $\delta_1 = 1.500$, $\delta_2 = 1.000$, $\delta_3 = 0.866$, and $\delta_4 = 0.003$. By Result 9.8, the eigenvalues of $\mathbf{X}^T\mathbf{X}$ are the squares of the singular values of **X**. Thus, $\lambda_1 = 2.250$, $\lambda_2 = 1.000$, $\lambda_3 = 0.750$, and $\lambda_4 = 0.000$. Also using a computer, the corresponding eigenvectors of

$$\mathbf{X}^T\mathbf{X} = \begin{pmatrix} 1.000 & 0.798 & 0.003 & 0.783 \\ 0.798 & 1.000 & 0.000 & 0.250 \\ 0.003 & 0.000 & 1.000 & 0.000 \\ 0.783 & 0.250 & 0.000 & 1.000 \end{pmatrix}.$$

1 One example is the issue of *centering* **X**, namely, whether one should center the matrix **X** (by subtracting the mean of each column from each element of the column) before computing the condition indices.

are found to be

$$\mathbf{V} = \begin{pmatrix} 0.6667 & -0.0006 & -0.0047 & -0.7453 \\ 0.5303 & 0.0019 & -0.6996 & 0.4788 \\ 0.0016 & -1.0000 & 0.5658 & 0.0022 \\ 0.5237 & 0.0019 & 0.7145 & 0.4639 \end{pmatrix}.$$

Note that the numbers are rounded to the nearest four decimal places. We then compute the condition indices as follows:

$$\kappa_1 = \frac{\delta_1}{\delta_1} = \frac{1.5}{1.5} = 1 \text{ (as expected)}, \qquad \kappa_2 = \frac{\delta_1}{\delta_2} = \frac{1.5}{1.0} = 1.5,$$

$$\kappa_3 = \frac{\delta_1}{\delta_3} = \frac{1.5}{0.866} = 1.7 \qquad \kappa_4 = \frac{\delta_1}{\delta_4} = \frac{1.5}{0.003} = 500.^{2}$$

Three of the condition indices are very small and one is very large. We conclude then that $\mathbf{X}$ contains only one set of collinearity. The fourth condition number of $\mathbf{X}$, κ_4, is much larger than 100, indicating an extreme collinearity. This large value of κ_4 indicates that δ_4 is too small relative to the largest value, δ_1. Thus, δ_4 can be considered approximately equal to 0. Then by Result 9.8(c), one would expect that $\mathbf{X}\mathbf{v}_4 \cong \mathbf{0}$, where $\cong$ indicates approximate equality and $\mathbf{v}_4$ is the fourth column of the eigenvector matrix $\mathbf{V}$ above. Indeed,

$$\mathbf{X}\mathbf{v}_4 = \begin{pmatrix} 0.322 & 0.5 & -0.5 & 0.0 \\ 0.012 & 0.5 & 0.5 & -0.5 \\ 0.631 & 0.5 & 0.0 & 0.5 \\ 0.316 & 0.0 & 0.5 & 0.5 \\ 0.000 & 0.0 & 0.5 & 0.0 \\ 0.631 & 0.5 & 0.0 & 0.5 \end{pmatrix} \begin{pmatrix} -0.7453 \\ 0.4788 \\ 0.0022 \\ 0.4639 \end{pmatrix} = \begin{pmatrix} -0.0017 \\ -0.0004 \\ 0.0010 \\ -0.0025 \\ 0.0011 \\ 0.0010 \end{pmatrix} \cong \mathbf{0}.$$

Now, writing $\mathbf{X}$ in terms of its columns as $\mathbf{X} = (\mathbf{x}_1 : \mathbf{x}_2 : \mathbf{x}_3 : \mathbf{x}_4)$, we have

$$\mathbf{X}\mathbf{v}_4 = (\mathbf{x}_1 : \mathbf{x}_2 : \mathbf{x}_3 : \mathbf{x}_4) \begin{pmatrix} -0.7453 \\ 0.4788 \\ 0.0022 \\ 0.4639 \end{pmatrix}$$

$$= -0.7453\ \mathbf{x}_1 + 0.4788\ \mathbf{x}_2 + 0.0022\ \mathbf{x}_3 + 0.4639\ \mathbf{x}_4. \tag{11.10}$$

Equating (11.10) to $\mathbf{0}$ and judging the coefficient of $\mathbf{x}_3$ to be close enough to 0, we

2 Note that if you use a computer to compute the singular values of $\mathbf{X}$, you will obtain $\kappa_4 =$ 493.9 not 500. This is because in our calculations we rounded the singular values to the nearest three decimal places.

obtain

$$-0.7453\ \mathbf{x}_1 + 0.4788\ \mathbf{x}_2 + 0.4639\ \mathbf{x}_4 = \mathbf{0},$$

which implies

$$\mathbf{x}_1 = 0.64\ \mathbf{x}_2 + 0.62\ \mathbf{x}_4 \cong 0.63\ (\mathbf{x}_2 + \mathbf{x}_4\ );$$

that is, $\mathbf{x}_1$ is approximately proportional to the sum of $\mathbf{x}_2$ and $\mathbf{x}_4$. Note here that only three of the four variables are involved in the collinearity ($\mathbf{x}_3$, the third column of $\mathbf{X}$, is not involved).

It can be seen from the above example that the singular values of $\mathbf{X}$ or, equivalently, the eigenvalues and eigenvectors of $\mathbf{X}^T\mathbf{X}$ are effective in diagnosing collinearity. The eigenvalues and eigenvectors are also very useful in many other applications, as we shall see below.

11.3. Classification of Quadratic Forms

A *quadratic function* in one variable x_1 is given by

$$f(x_1) = ax_1^2 + bx_1 + c, \tag{11.11}$$

where a, b, and c are scalars and $a \neq 0$. The function in (11.11) is called a quadratic function because the highest power of x_1 is 2. A quadratic function is also called a *quadratic form*. If $a = 0$ but $b \neq 0$, then (11.11) becomes a linear function (again because the highest power of x_1 is 1). Equation (11.11) can be generalized to two or more variables. For example, the equation

$$q(x_1, x_2) = 6x_1^2 + 4x_1x_2 + 9x_2^2 \tag{11.12}$$

is a quadratic function in x_1 and x_2. Equation (11.12) can be written in matrix notation as

$$q(x_1, x_2) = \begin{pmatrix} x_1 & x_2 \end{pmatrix} \begin{pmatrix} 6 & 2 \\ 2 & 9 \end{pmatrix} \begin{pmatrix} x_1 \\ x_2 \end{pmatrix} = \mathbf{x}^T\mathbf{A}\mathbf{x}, \tag{11.13}$$

where

$$\mathbf{x} = \begin{pmatrix} x_1 \\ x_2 \end{pmatrix} \quad \text{and} \quad \mathbf{A} = \begin{pmatrix} 6 & 2 \\ 2 & 9 \end{pmatrix}.$$

By carrying out the multiplication in (11.13), one can easily verify that $\mathbf{x}^T\mathbf{A}\mathbf{x} = q(x_1, x_2)$. We should note here that the matrix $\mathbf{A}$ is not unique. For example, one can also verify that $\mathbf{x}^T\mathbf{B}\mathbf{x} = q(x_1, x_2)$, where

$$\mathbf{B} = \begin{pmatrix} 6 & 1 \\ 3 & 9 \end{pmatrix}.$$

However, any quadratic form can always be expressed in terms of a unique symmetric matrix. Thus, if $\mathbf{B}$ is asymmetric, one can show that $\mathbf{x}^T\mathbf{B}\mathbf{x} = \mathbf{x}^T\mathbf{A}\mathbf{x}$, where

$$\mathbf{A} = 0.5\,(\mathbf{B} + \mathbf{B}^{\mathrm{T}}).$$

Notice here that **A** is symmetric by Result 2.1(a). Without loss of generality, then, we take the matrix **A** to be symmetric. If **A** is symmetric, it is also unique.

To obtain (11.13) from (11.12), we create a vector **x** and a square matrix **A**. The vector **x** contains the variables. The coefficients of the squared variables in (11.12) are the diagonal elements of **A**. The coefficient of the cross-product term (4 in this case) is divided by 2, and the result is the corresponding off-diagonal elements of **A**.

More generally, if **x** is a $k \times 1$ vector and **A** is a $k \times k$ symmetric matrix, one can show that

$$q(\mathbf{x}) = \mathbf{x}^{\mathrm{T}}\mathbf{A}\mathbf{x} \qquad (11.14)$$

is a quadratic function of the variables in **x**. The function $\mathbf{x}^{\mathrm{T}}\mathbf{A}\mathbf{x}$ is known as a *quadratic form*. It is a quadratic form in many variables and the notation in (11.14) is clearly independent of the number of variables in **x**. Note also that $q(\mathbf{x})$ is a scalar function of **x** completely determined by the matrix **A**. In other words, the matrix **A** defines the quadratic form. Thus, the matrix **A** is referred to as the *matrix of the quadratic form*.

Quadratic forms are encountered in many practical applications. For example, we have already seen quadratic forms in (8.50), where the residual sum of squares is expressed as $\mathbf{e}^{\mathrm{T}}\mathbf{e} = \mathbf{y}^{\mathrm{T}}(\mathbf{I}_n - \mathbf{P})\mathbf{y}$. Here, $\mathbf{x} = \mathbf{y}$ and $\mathbf{A} = (\mathbf{I}_n - \mathbf{P})$. Expressing the residual sum of squares in this form facilitates the derivation of several theoretical properties of the residuals and related quantities in regression analysis. Additionally, as we shall see later in this section, quadratic forms are useful, for example, for comparing matrices and for the classification of symmetric matrices into several types.

As we mentioned above, the matrix of a quadratic form can always be expressed in terms of a unique symmetric matrix. Therefore, in the sequel, we focus our attention on real symmetric matrices.

Clearly, if $\mathbf{x} = \mathbf{0}$, then $q(\mathbf{x}) = 0$ for all **A**. Quadratic forms are classified according to the sign of $q(\mathbf{x})$ for all values of **x** other than $\mathbf{x} = \mathbf{0}$, as given in the following definitions.

Definitions. A quadratic form $q(\mathbf{x}) = \mathbf{x}^{\mathrm{T}}\mathbf{A}\mathbf{x}$ is said to be:
(a) *positive definite* if $q(\mathbf{x}) > 0$ for all **x** other than $\mathbf{x} = \mathbf{0}$.
(b) *positive semidefinite* if $q(\mathbf{x}) \geq 0$ for all **x**, but $q(\mathbf{x}) = 0$ for some $\mathbf{x} \neq \mathbf{0}$.
(c) *negative definite* if $q(\mathbf{x}) < 0$ for all **x** other than $\mathbf{x} = \mathbf{0}$.
(d) *negative semidefinite* if $q(\mathbf{x}) \leq 0$ for all **x**, but $q(\mathbf{x}) = 0$ for some $\mathbf{x} \neq \mathbf{0}$.
(e) *indefinite* otherwise.

Classes (a) and (b) together are called *nonnegative definite*, whereas classes (c) and (d) together are called *nonpositive definite*. Now, given a quadratic form, how can we tell to which class it belongs? It is clearly difficult to use the above definitions to classify a particular quadratic form. Fortunately, the above classification is also determined by the eigenvalues of **A** as stated in the next result.

Result 11.2. Let **A** be the symmetric matrix of a quadratic form. The quadratic form and the matrix **A** itself are said to be:

(a) *positive definite* if all of the eigenvalues of **A** are positive.
(b) *positive semidefinite* if all of the eigenvalues of **A** are nonnegative but some are 0.
(c) *negative definite* if all of the eigenvalues of **A** are negative.
(d) *negative semidefinite* if all of the eigenvalues of **A** are nonpositive but some are 0.
(e) *indefinite* if some of the eigenvalues of **A** are positive and some are negative.

Thus, for example, $q(x_1, x_2)$ in (11.12) is positive definite because the corresponding quadratic form matrix **A** in (11.13) has eigenvalues of 10 and 5, both positive. (Note that **A** in (11.13) is the same as **A** in (9.4) for which we have already computed the eigenvalues.) Therefore, $q(x_1, x_2) > 0$ for all values of x_1 and x_2, other than $x_1 = 0$ and $x_2 = 0$. On the other hand, the quadratic form

$$2x_1^2 + 4x_1 x_2 + 2x_2^2 = \begin{pmatrix} x_1 & x_2 \end{pmatrix} \begin{pmatrix} 2 & 2 \\ 2 & 2 \end{pmatrix} \begin{pmatrix} x_1 \\ x_2 \end{pmatrix}$$

is positive semidefinite because the matrix of the quadratic form can be shown to have eigenvalues of 4 and 0.

By contrasting (11.14) with (10.16), we see that when **A** is positive definite and we set $q(\mathbf{x})$ to a positive constant, the quadratic form in (11.14) is an equation of an ellipsoid centered at the origin.

Matrix Inequality. In Chapter 1, we defined the matrix equality $\mathbf{A} = \mathbf{B}$ to mean that **A** and **B** have the same order and that $a_{ij} = b_{ij}$ for all values of i and j. We now define matrix inequality. For any comparison between two matrices to be meaningful, the two matrices must be of the same order. There is no unique way to compare matrices in terms of their magnitudes; there are at least three different ways to do this. The first and least common way is based on element-wise comparison of the two matrices. Accordingly, for any two matrices **A** and **B** of the same order, if $a_{ij} > b_{ij}$, for all i and j, we say that $\mathbf{A} > \mathbf{B}$.

The second and most common definition is based on quadratic forms and is useful for comparing square matrices of the same order. It states that

$$\mathbf{A} > \mathbf{B} \text{ if and only if } \mathbf{x}^{\mathrm{T}}\mathbf{A}\mathbf{x} > \mathbf{x}^{\mathrm{T}}\mathbf{B}\mathbf{x}, \text{ for all vectors } \mathbf{x}. \qquad (11.15)$$

An equivalent way of stating (11.15) is to say

$$\mathbf{A} > \mathbf{B} \text{ if and only if } \mathbf{A} - \mathbf{B} \text{ is positive definite.} \qquad (11.16)$$

The third definition is based on matrix norms. It is valid for any matrices of the same order. It states that

$$\mathbf{A} > \mathbf{B} \text{ if and only if } \|\mathbf{A}\|_p > \|\mathbf{B}\|_p, \qquad (11.17)$$

for a given value of p (usually 2).

Quadratic forms are frequently encountered in many applications, such as multivariate and regression analysis. In these applications, it is usually of interest to classify a particular quadratic form and its associated symmetric matrix **A** as in Result 11.2. In this section, we have seen that the eigenvalues of **A** make this classification an easy task.

*11.4. Markov Chains

Markov chains are examples of oblique transformations. Suppose that a person can be affiliated with one of three political parties. During a specified period of time, say a year, a person who belongs to a given party can either remain in the same party or switch to one of the other two parties. The probabilities of these events are given in the following matrix:

$$\mathbf{A} = \begin{pmatrix} 0.4 & 0.2 & 0.2 \\ 0.4 & 0.7 & 0.2 \\ 0.2 & 0.1 & 0.6 \end{pmatrix}. \tag{11.18}$$

The element a_{ij} represents the probability that a person who is currently in party j is going to switch to party i during the next year. Clearly, the sum of each column in **A** must be 1. The matrix **A** is called a *transition matrix*. Now, let $\mathbf{x}_t$ be a three-element vector containing the percentages of people who belong to each of the three parties at the end of year t. Currently (i.e., at $t = 0$), the percentages of people who belong to the three parties are 50 percent, 30 percent, and 20 percent, respectively. Therefore, the initial probability distribution is given by

$$\mathbf{x}_0 = \begin{pmatrix} 0.5 \\ 0.3 \\ 0.2 \end{pmatrix}. \tag{11.19}$$

Suppose further that the relationship between $\mathbf{x}_{t+1}$ and $\mathbf{x}_t$ is given by

$$\mathbf{x}_{t+1} = \mathbf{A}\mathbf{x}_t. \tag{11.20}$$

Thus, for example,

$$\mathbf{x}_1 = \mathbf{A}\mathbf{x}_0 \tag{11.21}$$

$$= \begin{pmatrix} 0.4 & 0.2 & 0.2 \\ 0.4 & 0.7 & 0.2 \\ 0.2 & 0.1 & 0.6 \end{pmatrix} \begin{pmatrix} 0.5 \\ 0.3 \\ 0.2 \end{pmatrix} = \begin{pmatrix} 0.30 \\ 0.45 \\ 0.25 \end{pmatrix},$$

from which it follows that at the end of the first year (the beginning of the second year) the percentages of people who belong to the three parties are 30 percent, 45 percent, and 25 percent, respectively.

By comparison with (7.25), it can be seen that (11.20) is a linear transformation, that is, $\mathbf{x}_{t+1}$ is a linear transformation of $\mathbf{x}_t$. But since the matrix **A** in (11.18) is not orthogonal, the transformation is oblique, that is, the axes are obliquely rotated.

The process just described is known as the *Markov process* or *Markov chain*. Two practical questions arise in this context. The first is: Starting at $t = 0$ and given the transition matrix **A** and the initial distribution $\mathbf{x}_0$, what is the distribution at the end of the nth year? For example, the distribution at the end of the second year is

$$\mathbf{x}_2 = \mathbf{A}\mathbf{x}_1 = \mathbf{A}\mathbf{A}\mathbf{x}_0 = \mathbf{A}^2\mathbf{x}_0$$

and the distribution at the end of the third year is

$$\mathbf{x}_3 = \mathbf{A}\mathbf{x}_2 = \mathbf{A}\mathbf{A}^2\mathbf{x}_0 = \mathbf{A}^3\mathbf{x}_0.$$

Continuing in this manner, the distribution at the end of the nth year is found to be

$$\mathbf{x}_n = \mathbf{A}^n\mathbf{x}_0. \tag{11.22}$$

Thus, to find the distribution at the end of the nth year, we need to multiply $\mathbf{A}$ by itself n times and then post-multiply the product by $\mathbf{x}_0$. For a large value of n, this can be a computationally tedious process. However, if $\mathbf{A}$ has a linearly independent set of eigenvectors, Result 9.2(b) can be used to simplify the computation of $\mathbf{A}^n$ considerably. The eigenvalues and eigenvectors of $\mathbf{A}$ in (11.18) are found to be

$$\Lambda = \begin{pmatrix} 1 & 0 & 0 \\ 0 & 0.5 & 0 \\ 0 & 0 & 0.2 \end{pmatrix} \quad \text{and} \quad \mathbf{V} = \begin{pmatrix} 0.418 & 0.000 & -0.802 \\ 0.837 & -0.745 & 0.535 \\ 0.418 & 0.745 & 0.267 \end{pmatrix}.$$

The matrix $\mathbf{V}$ is nonsingular and its inverse is

$$\mathbf{V}^{-1} = \begin{pmatrix} 0.598 & 0.598 & 0.598 \\ 0.000 & -0.447 & 0.894 \\ -0.935 & 0.312 & 0.312 \end{pmatrix}.$$

By Result 9.2(b), we have

$$\begin{aligned} \mathbf{A}^n &= \mathbf{V}\Lambda^n\mathbf{V}^{-1} \\ &= \begin{pmatrix} 0.418 & 0.000 & -0.802 \\ 0.837 & -0.745 & 0.535 \\ 0.418 & 0.745 & 0.267 \end{pmatrix} \begin{pmatrix} 1^n & 0 & 0 \\ 0 & 0.5^n & 0 \\ 0 & 0 & 0.2^n \end{pmatrix} \begin{pmatrix} 0.598 & 0.598 & 0.598 \\ 0.000 & -0.447 & 0.894 \\ -0.935 & 0.312 & 0.312 \end{pmatrix}. \end{aligned}$$

For example, for $n = 5$, we have

$$\begin{aligned} \mathbf{A}^5 &= \begin{pmatrix} 0.418 & 0.000 & -0.802 \\ 0.837 & -0.745 & 0.535 \\ 0.418 & 0.745 & 0.267 \end{pmatrix} \begin{pmatrix} 1^5 & 0 & 0 \\ 0 & 0.5^5 & 0 \\ 0 & 0 & 0.2^5 \end{pmatrix} \begin{pmatrix} 0.598 & 0.598 & 0.598 \\ 0.000 & -0.447 & 0.894 \\ -0.935 & 0.312 & 0.312 \end{pmatrix} \\ &= \begin{pmatrix} 0.250 & 0.250 & 0.250 \\ 0.500 & 0.510 & 0.479 \\ 0.250 & 0.240 & 0.271 \end{pmatrix}, \end{aligned}$$

from which it follows that

$$\mathbf{x}_5 = \mathbf{A}^5\mathbf{x}_0 = \begin{pmatrix} 0.250 & 0.250 & 0.250 \\ 0.500 & 0.510 & 0.479 \\ 0.250 & 0.240 & 0.271 \end{pmatrix} \begin{pmatrix} 0.5 \\ 0.3 \\ 0.2 \end{pmatrix} = \begin{pmatrix} 0.250 \\ 0.499 \\ 0.251 \end{pmatrix}.$$

Therefore, at the end of the fifth year the percentages of people who belong to the three parties are 25 percent, 49.9 percent, and 25.1 percent, respectively.

The second question is: What will be the distribution in the very long run, i.e., when n goes to infinity? In this case we have

$$\Lambda^\infty = \begin{pmatrix} 1^\infty & 0 & 0 \\ 0 & 0.5^\infty & 0 \\ 0 & 0 & 0.2^\infty \end{pmatrix} = \begin{pmatrix} 1 & 0 & 0 \\ 0 & 0 & 0 \\ 0 & 0 & 0 \end{pmatrix},$$

from which it follows that

$$\mathbf{A}^\infty = \begin{pmatrix} 0.418 & 0.000 & -0.802 \\ 0.837 & -0.745 & 0.535 \\ 0.418 & 0.745 & 0.267 \end{pmatrix} \begin{pmatrix} 1 & 0 & 0 \\ 0 & 0 & 0 \\ 0 & 0 & 0 \end{pmatrix} \begin{pmatrix} 0.598 & 0.598 & 0.598 \\ 0.000 & -0.447 & 0.894 \\ -0.935 & 0.312 & 0.312 \end{pmatrix}$$

$$= \begin{pmatrix} 0.25 & 0.25 & 0.25 \\ 0.50 & 0.50 & 0.50 \\ 0.25 & 0.25 & 0.25 \end{pmatrix},$$

hence

$$\mathbf{x}_\infty = \mathbf{A}^\infty\mathbf{x}_0 = \begin{pmatrix} 0.25 & 0.25 & 0.25 \\ 0.50 & 0.50 & 0.50 \\ 0.25 & 0.25 & 0.25 \end{pmatrix} \begin{pmatrix} 0.5 \\ 0.3 \\ 0.2 \end{pmatrix} = \begin{pmatrix} 0.25 \\ 0.50 \\ 0.25 \end{pmatrix}. \tag{11.23}$$

Therefore, in the very long run the percentages of people who belong to the three parties will be 25 percent, 50 percent, and 25 percent, respectively. Compare this with the initial distribution in (11.19).

The distribution in (11.23) is referred to as the *steady-state distribution*. In fact, one property of Markov chains is that the steady-state distribution is determined solely by the transition matrix $\mathbf{A}$; that is, it is independent of the initial distribution vector $\mathbf{x}_0$. To see this, suppose $\mathbf{x}_0$ in (11.19) is replaced by

$$\mathbf{x}_0 = \begin{pmatrix} p \\ q \\ 1-p-q \end{pmatrix},$$

where $0 \le p \le 1$ and $0 \le q \le 1$, so that the sum of the elements of $\mathbf{x}_0$ is 1. We now have

$$\mathbf{x}_\infty = \mathbf{A}^\infty \mathbf{x}_0 = \begin{pmatrix} 0.25 & 0.25 & 0.25 \\ 0.50 & 0.50 & 0.50 \\ 0.25 & 0.25 & 0.25 \end{pmatrix} \begin{pmatrix} p \\ q \\ 1-p-q \end{pmatrix} = \begin{pmatrix} 0.25 \\ 0.50 \\ 0.25 \end{pmatrix},$$

which is the same as (11.23). Therefore, we conclude that the steady-state distribution is independent of the initial distribution. For other properties of Markov chains and other types of stochastic (probabilistic) processes, the interested reader is referred to books such as Karlin and Taylor (1975), Kannan (1979), or Bhattacharya and Waymire (1990).

*11.5. Principal Component Analysis

Principal component analysis (PCA) is a multivariate analysis technique in which we start with an $n \times p$ data matrix $\mathbf{X}$. The columns of $\mathbf{X}$ represent p variables of interest and the rows represent n observations. The variables are usually correlated, and one objective of PCA is to transform the data into p orthogonal variables, the first k of which contain almost all the information contained in the original p variables. The new set of k variables is known as the *principal components* (PC) of $\mathbf{X}$. The p original variables can then be replaced by the (hopefully much) smaller set of k PC.

Because we start with p variables and end up with $k < p$ principal components, we may think of PCA as a dimension reduction technique. Also, because the original variables are usually correlated but the PC are orthogonal, PCA is also a transformation to achieve linear independence.

A starting point in PCA is to compute the variance-covariance matrix of the variables. This matrix is defined in (8.28) as

$$\mathbf{S} = (n-1)^{-1}\, \mathbf{X}^{\mathrm{T}}\mathbf{C}\mathbf{X}, \tag{11.24}$$

where $\mathbf{C}$ is the centering matrix defined in (8.26). We next compute the eigenvalues and eigenvectors of $\mathbf{S}$ and store them in the matrices Λ and $\mathbf{V}$ as in (9.11). Using Result 9.3(d), the matrix $\mathbf{S}$ can be decomposed into

$$\mathbf{S} = \mathbf{V}\Lambda\mathbf{V}^{\mathrm{T}}. \tag{11.25}$$

Finally, we compute

$$\mathbf{W} = \mathbf{X}\mathbf{V}. \tag{11.26}$$

The columns of this $n \times p$ matrix $\mathbf{W}$ are the PC of $\mathbf{X}$. Note that if we transpose both sides of (11.26), we obtain $\mathbf{W}^{\mathrm{T}} = \mathbf{V}^{\mathrm{T}}\mathbf{X}^{\mathrm{T}}$, which implies that each row of $\mathbf{W}$ is an orthogonal linear transformation of the corresponding row of $\mathbf{X}$, because the matrix of transformation $\mathbf{V}^{\mathrm{T}}$ is an orthogonal matrix.

Now, let us compute the variance-covariance matrix of $\mathbf{W}$. By analogy with (11.24), the variance-covariance matrix of $\mathbf{W}$ is

$$\mathbf{A} = (n-1)^{-1}\, \mathbf{W}^{\mathrm{T}}\mathbf{C}\mathbf{W}. \tag{11.27}$$

Substituting (11.26), and noting that $\mathbf{W}^T = \mathbf{V}^T\mathbf{X}^T$, we obtain

$$\mathbf{A} = (n-1)^{-1}\, \mathbf{V}^T\mathbf{X}^T\mathbf{C}\mathbf{X}\mathbf{V} = \mathbf{V}^T\mathbf{S}\mathbf{V},$$

where $\mathbf{S}$ is the variance-covariance matrix of $\mathbf{X}$ given by (11.24). We finally substitute (11.25) for $\mathbf{S}$ and obtain

$$\mathbf{A} = \mathbf{V}^T\mathbf{V}\Lambda\mathbf{V}^T\mathbf{V} = \Lambda, \tag{11.28}$$

because, by Result 9.3(c), $\mathbf{V}^T\mathbf{V} = \mathbf{I}$. Recall that Λ is a diagonal matrix whose diagonal elements are the ordered eigenvalues of $\mathbf{S}$ and, by Result 9.4, all of the eigenvalues of $\mathbf{S}$ are nonnegative. It follows then that the variance of the jth principal component is λ_j, the jth largest eigenvalue of $\mathbf{S}$. Furthermore, the covariance between any two columns of $\mathbf{W}$ is 0. Therefore the PC are indeed orthogonal.

So far we have achieved the first objective, that is, to obtain p orthogonal variables. The next objective is to reduce the number of PC without much loss of the information contained in $\mathbf{X}$. To achieve this objective, we make use of the following result.

Result 11.3. Let $\mathbf{S}$ and $\mathbf{A}$ be as defined in (11.24) and (11.27); then $\mathrm{tr}(\mathbf{S}) = \mathrm{tr}(\Lambda)$.

In words, the sum of the variances of the original variables in $\mathbf{X}$ is equal to the sum of the variances of the PC in $\mathbf{W}$. By (11.28), this total variance is the sum of the eigenvalues of $\mathbf{S}$. Since the eigenvalues are arranged in descending order, it may happen that the first k of them account for most of the total variance. If this is the case, the first k PC contain almost all of the total variance of $\mathbf{X}$. We can then replace the p variables in $\mathbf{X}$ by the first k PC and thereby achieve the second objective.

It is now time for an illustrative example. Let us consider the weights and heights in the 9×2 data matrix $\mathbf{X}$ in (10.45). Because this is a small data set, the benefits of PCA as a dimension reduction method will not be really felt or appreciated. This matrix, however, will illustrate the transformation to achieve orthogonality. The variance-covariance matrix of $\mathbf{X}$ is

$$\mathbf{S} = \begin{pmatrix} 12.750 & 9.625 \\ 9.625 & 8.250 \end{pmatrix}.$$

The eigenvalues and eigenvectors of $\mathbf{S}$ are found to be

$$\Lambda = \begin{pmatrix} 20.384 & 0 \\ 0 & 0.616 \end{pmatrix}$$

and

$$\mathbf{V} = \begin{pmatrix} 0.783 & -0.621 \\ 0.621 & 0.783 \end{pmatrix},$$

to the nearest three decimal places. We then compute the PC as defined in (11.26) and obtain

$$\mathbf{W} = \mathbf{XV} = \begin{pmatrix} 186.69 & -51.07 \\ 190.74 & -49.18 \\ 176.69 & -50.80 \\ 180.74 & -48.91 \\ 183.71 & -49.99 \\ 179.34 & -49.07 \\ 188.71 & -50.13 \\ 184.50 & -50.61 \\ 182.31 & -50.15 \end{pmatrix},$$

to the nearest two decimal places. The variance-covariance matrix of **W** is

$$\begin{pmatrix} 20.384 & 0 \\ 0 & 0.616 \end{pmatrix},$$

which can be seen to be equal to Λ. Thus, the PC are orthogonal, as expected. The scatter plot of the columns of **W** is shown in Figure 11.1. Compared with the scatter plot of the columns of **X** given in Figure 10.8, we see that although there is a trend between the two columns of **X**, there is no trend at all between the columns of **W**.

Notice also that $\text{tr}(\mathbf{S}) = \text{tr}(\mathbf{A}) = \text{tr}(\Lambda)$, as guaranteed by Result 11.3. Since

$$\frac{\lambda_1}{\lambda_1 + \lambda_2} = \frac{20.384}{20.384 + 0.616} = 0.97,$$

then 97 percent of the total variability in **X** is accounted for by the first PC alone. Therefore, the two variables in **X** can be replaced by only the first principal component without much loss of information.

Note that principal component analysis can serve as an example of both linear transformations and the eigenvalue problem.

*11.6. Factor Analysis

Factor analysis is a multivariate analysis technique designed to handle large data sets with correlated variables. The relationships among the variables are difficult to understand and interpret. One objective of factor analysis is to explain these relationships in terms of a few constructed variables. These constructed variables are called *factors*. The factors should be easier to interpret than the original set of variables. The factors should also be as independent of each other as possible. Thus, factor analysis, like principal component analysis, can be thought of as a dimension reduction technique.

The factor analytic model with k factors can be described as

$$\mathbf{x} = \mu + \mathbf{Lf} + \varepsilon. \tag{11.29}$$

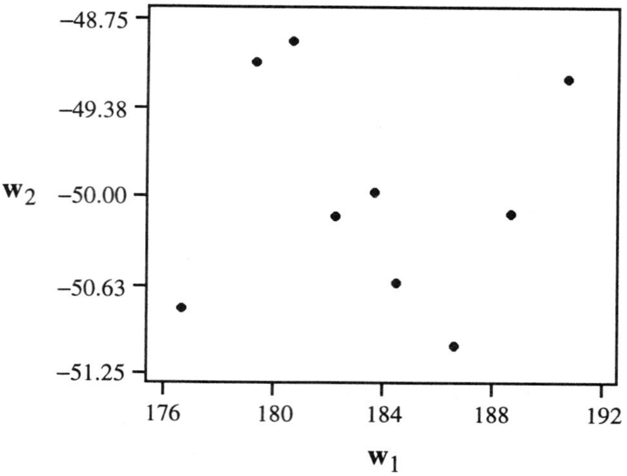

Figure 11.1. Scatter plot of the first PC, $\mathbf{w}_1$, versus the second PC, $\mathbf{w}_2$.

Here $\mathbf{x}$ is a $p \times 1$ vector representing the p original variables, $\boldsymbol{\mu}$ is the mean vector of $\mathbf{x}$, and $\mathbf{f}$ is a $k \times 1$ vector representing the k factors (constructed variables), where k is (hopefully much) smaller than p. The $p \times k$ matrix $\mathbf{L}$ is known as the matrix of *factor loadings* because its ijth element represents the contribution of the ith variable to the jth factor. If $\mathbf{l}_i^{\mathrm{T}}$ is the ith row of $\mathbf{L}$, the ith variable, x_i, can be written as

$$x_i = \mu_i + \mathbf{l}_i^{\mathrm{T}}\mathbf{f} + \varepsilon_i, \tag{11.30}$$

where μ_i is the ith element of $\boldsymbol{\mu}$ and ε_i is the ith element of $\boldsymbol{\varepsilon}$. Accordingly, the ith variable is equal to $\mu_i + \varepsilon_i$ plus a linear combination of the factors $\mathbf{f}$. Thus, the factor $\mathbf{f}$ is common to all variables, but ε_i is specific to the ith variable. For this reason $\mathbf{f}$ is known as the *common* factor and ε_i is known as the *specific* factor.

In the k-factor model (11.29) the only observable quantity is $\mathbf{x}$. All of the components on the right-hand side of (11.29) are unobservable quantities. If certain assumptions hold, these quantities can be estimated from the data. It is beyond the scope and purpose of this book to give even a brief presentation of factor analysis, e.g., to specify the assumptions or discuss the methods by which the above quantities can be estimated, or discuss the various methods of factor rotation. The interested reader is referred to books on multivariate analysis, e.g., Johnson and Wichern (1992). The point we wish to make here is that in factor analysis there is a need for linear transformations, both orthogonal and oblique.

Of all the quantities on the right-hand side of (11.29), the matrix of factor loadings $\mathbf{L}$ is of particular interest because the interpretability of the factors hinges on the structure of $\mathbf{L}$. If $\mathbf{L}$ has a simple structure, then the factors can be easily and meaningfully interpreted, but if the initial estimate does not have a simple structure, a meaningful interpretation might be difficult to obtain. However, there may be a way out of this problem, as explained below.

We know that, for any nonsingular $k \times k$ matrix $\mathbf{A}$, $\mathbf{A}^{-1}\mathbf{A} = \mathbf{I}$. Therefore, inserting $\mathbf{A}^{-1}\mathbf{A}$ between $\mathbf{L}$ and $\mathbf{f}$ in (11.29) will have no effect on the equation. The factor model in (11.29) can then be expressed as

$$\mathbf{x} = \mu + \mathbf{L}\mathbf{A}^{-1}\mathbf{A}\mathbf{f} + \varepsilon = \mu + \mathbf{L}^* \mathbf{f}^* + \varepsilon, \tag{11.31}$$

where $\mathbf{L}^* = \mathbf{L}\mathbf{A}^{-1}$ and $\mathbf{f}^* = \mathbf{A}\mathbf{f}$. Although (11.29) and (11.31) are different, they are equivalent representations of the k-factor model. In (11.31), $\mathbf{f}^*$ is a new set of factors which are related to the initial set of factors by a linear transformation. Similarly, $\mathbf{L}^*$ is a new matrix of factor loadings which is related to the initial matrix by a linear transformation. Therefore, we conclude that there is no unique representation of the k-factor model. For this reason, factor analysis is a highly controversial subject. This disadvantage, however, can sometimes be turned into an advantage, as explained below.

Recall that one of the objectives of factor analysis is to find factors that are conceptually meaningful and easy to interpret, and that the interpretability of the factors depends on the structure of the matrix of factor loadings $\mathbf{L}$. So, it may happen that the initial matrix $\mathbf{L}$ does not have a simple structure, but the transformed matrix $\mathbf{L}^*$ does. Since we can choose the matrix $\mathbf{A}$ in (11.31) in any way we like as long as it is nonsingular, we should choose $\mathbf{A}$ in such a way as to make $\mathbf{L}^*$ easy to interpret. From the statistical point of view, it is generally preferable to choose $\mathbf{A}$ to be an orthogonal matrix. The resultant transformation will then be an orthogonal rotation of the initial factors. However, an orthogonal transformation might not yield factors that are conceptually meaningful. In this case, we may have to choose a nonorthogonal matrix, and then the resultant transformation is an oblique rotation of the initial factors. This oblique rotation may result in a conceptually more meaningful set of factors. We wish to reiterate that many details have been omitted here because they are beyond the scope and purpose of this book.

Exercises

11.1. Let $\mathbf{A}$ be an $n \times m$ matrix and $\mathbf{1}_n$ be the vector of n ones. Show that the Frobenius norm of $\mathbf{A}$ can be expressed as

$$\|\mathbf{A}\|_F = +\sqrt{\mathrm{tr}(\mathbf{A}^{\mathrm{T}}\mathbf{A})} = +\sqrt{\mathrm{tr}(\mathbf{A}\mathbf{A}^{\mathrm{T}})} = +\sqrt{\mathbf{1}_n^{\mathrm{T}}(\mathbf{A}\cdot\mathbf{A})\mathbf{1}_m},$$

where $\mathbf{A}\cdot\mathbf{A}$ is the Hadamard product of $\mathbf{A}$ and itself.

11.2. Given $\mathbf{A} = \begin{pmatrix} 1 & 3 \\ 2 & 6 \end{pmatrix}$, $\mathbf{B} = \begin{pmatrix} 3 & 4 \\ 4 & -3 \end{pmatrix}$, and $\mathbf{C} = \begin{pmatrix} 2 & 1 & -2 \\ 1 & 0 & 1 \end{pmatrix}$:

(a) Compute the Frobenius norm of each matrix.
(b) Compute the p-norm of each matrix for $p = 1, 2, \infty$.
(c) Using each of the above norms, determine which matrix is the largest in magnitude. [Note, however, that in some applications it might not be meaningful to compare matrices of different orders.]

11.3. Using the matrices in Exercise 11.2, verify that:

(a) $\|-5\mathbf{A}\|_p = 5\,\|\mathbf{A}\|_p$, for $p = 1, 2, \infty$.

(b) $\|\mathbf{A} + \mathbf{B}\|_p \leq \|\mathbf{A}\|_p + \|\mathbf{B}\|_p$, for $p = 1, 2, \infty$.

(c) $\|\mathbf{AC}\|_p \leq \|\mathbf{A}\|_p \times \|\mathbf{C}\|_p$, for $p = 1, 2, \infty$.

11.4. Compute and comment on the condition indices for each of the matrices below.

$$\mathbf{X} = \begin{pmatrix} 2 & 4 \\ 1 & 2 \\ 1 & 1 \end{pmatrix}, \quad \mathbf{Y} = \begin{pmatrix} 1 & -1 \\ 0 & 1 \\ 1 & 1 \end{pmatrix}, \quad \text{and} \quad \mathbf{Z} = \begin{pmatrix} 4 & 0.000 \\ 0 & 0.001 \\ 3 & 0.000 \end{pmatrix}.$$

11.5. The row vector $\mathbf{r}$ below contains the eigenvalues of $\mathbf{X}^T\mathbf{X}$ for a given matrix $\mathbf{X}$, and the corresponding eigenvectors are the columns of $\mathbf{V}$:

$$\mathbf{r} = \begin{pmatrix} 3.2761 & 1.7229 & 0.9954 & 0.0048 & 0.0006 & 0.0002 \end{pmatrix}$$

$$\mathbf{V} = \begin{pmatrix} -0.29 & -0.65 & -0.03 & -0.59 & -0.38 & 0.09 \\ -0.34 & 0.32 & 0.66 & -0.42 & 0.23 & -0.33 \\ -0.39 & 0.36 & -0.52 & -0.01 & -0.39 & -0.54 \\ -0.44 & -0.34 & 0.38 & 0.69 & -0.21 & -0.14 \\ -0.49 & -0.22 & -0.38 & 0.02 & 0.75 & 0.03 \\ -0.46 & 0.43 & 0.00 & 0.01 & -0.20 & 0.75 \end{pmatrix}.$$

(a) Compute the condition indices of the matrix $\mathbf{X}$.

(b) How many sets of collinearities are there?

(c) Which variables are involved in each set?

(d) What are the exact or approximate relationships among the variables involved in each set of collinearity?

11.6. Let $\mathbf{A} = \begin{pmatrix} c & 0 \\ 0 & c \end{pmatrix}$ and $\mathbf{B} = \begin{pmatrix} 1+c & 1 \\ 1 & 1+c \end{pmatrix}$.

(a) Show that the eigenvalues of $\mathbf{A}$ are c and c (c of multiplicity 2).

(b) Show that the eigenvalues of $\mathbf{B}$ are $2 + c$ and c.

(c) Compute the condition numbers of $\mathbf{A}$ and $\mathbf{B}$.

(d) What can you conclude about the conditions of the matrices $\mathbf{A}$ and $\mathbf{B}$ as c approaches 0?

11.7. Using a computer, determine whether each of the data matrices

$$\mathbf{X} = \begin{pmatrix} 46 & 22 & 33 & 69 & 44 \\ 25 & -8 & -13 & 15 & -17 \\ -35 & 10 & 14 & -25 & 19 \\ 5 & 34 & -17 & 38 & 55 \\ -4 & -18 & -12 & 20 & -21 \\ 3 & 20 & 15 & 23 & 33 \end{pmatrix}, \quad \mathbf{Y} = \begin{pmatrix} 1.07 & -1.70 & 0.14 & 0.10 \\ 0.72 & 0.59 & -1.11 & 0.74 \\ 1.28 & 1.11 & 0.79 & 0.16 \\ 1.78 & -1.13 & -1.05 & -0.47 \\ 2.35 & -0.60 & 0.94 & 0.37 \\ 2.00 & 1.70 & -0.36 & -0.24 \end{pmatrix}$$

contains collinear columns. Specifically:

(a) Scale the columns of the data, if necessary.

(b) Compute the condition indices.

(c) If collinearity occurs, how many sets of collinearities are there, which variables are involved in each set, and how are they related?

11.8. Determine whether the matrices **A** and **B** in Exercise 9.3 and their corresponding quadratic forms are positive definite.

11.9. Determine whether the quadratic forms defined by the asymmetric matrices **C** and **D** in Exercise 9.3 are positive definite. [Hint: What are the corresponding symmetric matrices that define the same quadratic forms?]

11.10. Classify each of the quadratic forms in Exercise 9.1 according to type.

11.11. Classify each of the quadratic forms in Exercise 9.3 according to type.

11.12. Given $\mathbf{A} = \begin{pmatrix} a & 1 \\ 1 & b \end{pmatrix}$, find the conditions under which **A** is:

(a) positive definite.
(b) positive semidefinite.
(c) negative definite.
(d) indefinite.

11.13. Given the matrices **A** and **B** in Exercise 11.2, determine whether or not $\mathbf{A} > \mathbf{B}$ in the sense of (11.16).

11.14. Given the matrices **A** and **B** in Exercise 11.6, determine whether or not $\mathbf{A} > \mathbf{B}$ in the sense of (11.16).

11.15. There are only two competing supermarkets in a small town. Their shares of the market are currently 40 percent and 60 percent, respectively. To keep things simple, suppose that each family in the town buys groceries once every week and from only one of the two supermarkets. If a family goes to the first supermarket one week, there is a 30 percent chance that the family will go to the second supermarket the next week, whereas if the family goes to the second supermarket one week, there is a 50 percent chance that the family will go to the first supermarket the next week. What is the market share

(a) after one week?
(b) after four weeks?
(c) after one year?
(d) in the very long run?

11.16. Compute the steady-state distribution for each of the following transition matrices:

$$\mathbf{A} = \begin{pmatrix} 0.4 & 0.3 \\ 0.6 & 0.7 \end{pmatrix}, \qquad \mathbf{B} = \begin{pmatrix} 0.4 & 0.7 \\ 0.6 & 0.3 \end{pmatrix},$$

$$\mathbf{C} = \begin{pmatrix} 0.5 & 0.1 & 0.2 \\ 0.0 & 0.5 & 0.0 \\ 0.5 & 0.4 & 0.8 \end{pmatrix}, \qquad \mathbf{D} = \begin{pmatrix} 1.0 & 0.5 & 0.2 \\ 0.0 & 0.5 & 0.0 \\ 0.0 & 0.0 & 0.8 \end{pmatrix}.$$

11.17. Give the conditions on p and q so that $\mathbf{A}$ is a transition matrix, where:

(a) $\mathbf{A} = \begin{pmatrix} p & 1-q \\ 1-p & q \end{pmatrix}$. (b) $\mathbf{A} = \begin{pmatrix} p & 1-p \\ 1-q & q \end{pmatrix}$. (c) $\mathbf{A} = \begin{pmatrix} 0.5 & q \\ p & 1 \end{pmatrix}$.

11.18. Under the appropriate conditions on p and q in Exercise 11.17(a), compute the eigenvalues of $\mathbf{A}$ as a function of p and q.

11.19. Show that the least squares regression line (see Section 8.3), for the data in Figure 11.1, has a slope of 0.

11.20. For the data matrix $\mathbf{X}$ in Exercise 11.7, using a computer:
(a) Compute $\mathbf{S}$, the variance-covariance matrix of the data.
(b) Compute the spectral decomposition of $\mathbf{S}$ as in (11.25).
(c) Compute the principal components $\mathbf{W}$ as in (11.26).
Then answer the following questions.
(d) What would you expect pair-wise scatter plots of the columns of $\mathbf{W}$ to look like?
(e) Without any additional computations, what is the variance-covariance matrix of $\mathbf{W}$?
(f) What proportion of the total variability in the data is accounted for by the first column of $\mathbf{W}$?
(g) What proportion of the total variability in the data is accounted for by the first two columns of $\mathbf{W}$?
(h) How many of the principal components would capture most of the variability in the data?

11.21. Repeat Exercise 11.20 using the data matrix $\mathbf{Y}$ in Exercise 11.7.

Bibliography

Albert, A. (1972), *Regression and the Moore-Penrose Inverse*, New York: Academic Press.

Belsley, D. A. (1991), *Conditioning Diagnostics: Collinearity and Weak Data in Regression*, New York: John Wiley & Sons.

Belsley, D. A., Kuh, E. and Welsch, R. E. (1980), *Regression Diagnostics: Identifying Influential Data and Sources of Collinearity*, New York: John Wiley & Sons.

Bhattacharya, R. N. and Waymire, E. C. (1990), *Stochastic Processes with Applications*, New York: John Wiley & Sons.

Castillo, E., Gutiérrez, J. M. and Hadi, A. S. (1996), *Expert Systems and Probabilistic Network Models.*

Chatterjee, S. and Hadi, A. S. (1988), *Sensitivity Analysis in Linear Regression*, New York: John Wiley & Sons.

Chatterjee, S. and Price, B. (1991), *Regression Analysis by Example*, 2nd ed., New York: John Wiley & Sons.

Draper, N. R. and Smith, H. (1981), *Applied Regression Analysis*, 2nd ed., New York: John Wiley & Sons.

Fox, J. (1984), *Linear Statistical Models and Related Methods: With Applications to Social Research*, New York: John Wiley & Sons.

Golub, G. H. and Van Loan, C. (1989), *Matrix Computations*, 2nd ed., Baltimore: John Hopkins.

Gould, W. and Hadi, A. S. (1993), "Identifying Multivariate Outliers," *Stata Technical Bulletin*, 11, 2-5.

Graybill, F. A. (1983), *Matrices with Applications in Statistics*, 2nd ed., Belmont, Calif.: Wadsworth.

Hadi, A. S. (1986), "The Prediction Matrix: Its Properties and Role in Data Analysis," *Proceedings of the Business and Economic Statistics Section*, Washington, D.C.: The American Statistical Association, 631-636.

Hadi, A. S. (1992), "Identifying Multiple Outliers in Multivariate Data," *Journal of the Royal Statistical Society* (*B*), 54, 761-771.

Hadi, A. S. (1994), "A Modification of a Method for the Detection of Outliers in Multivariate Samples," *Journal of the Royal Statistical Society* (*B*), 56, 393-396.

Hadi, A. S. and Simonoff, J. S. (1993), "Procedures for the Identification of Multiple Outliers in Linear Models," *Journal of the American Statistical Association*, 88, 1264-1272.

Hadi, A. S. and Simonoff, J. S. (1995), "Improving the Estimation and Outlier Identification Properties of the Least Median of Squares and Minimum Volume Ellipsoid Estimators," *Parisankhyan Samikkha*, 1,61-70.

Hadi, A. S. and Wells, M. T. (1990), "Assessing the Effects of Multiple Rows on the Condition Number of a Matrix," *Journal of the American Statistical Association*, 85, 786-792.

Hartigan, J. (1975), *Clustering Algorithms*, New York: John Wiley & Sons.

Johnson, R. A. and Wichern, D. W. (1992), *Applied Multivariate Statistical Analysis*, 3rd ed., Englewood Cliffs, N.J.: Prentice-Hall.

Joseph, L. and Styan, G. P. H. (1987), "Selected Problems in Applied Matrix Algebra: Basic Matrix Algebra, Leontief Models, Linear Programming, & Least Squares," Technical Report, Department of Mathematics and Statistics, McGill University.

Kannan, D. (1979), *An Introduction to Stochastic Processes*, New York: Elsevier North Holland.

Karlin, S. and Taylor, H. M. (1975), *A First Course in Stochastic Processes*, 2nd ed., New York: Academic Press.

Kaufman, L. and Rousseeuw, P. J. (1990), *Finding Groups in Data: An Introduction to Cluster Analysis*, New York: John Wiley & Sons.

Neter, J., Wasserman, W. and Kutner, M. H. (1985), *Applied Regression Models*, 2nd ed., Homewood, Ill.: Irwin.

Noble, B. and Daniel, J. W. (1977), *Applied Linear Algebra*, 2nd ed., Englewood Cliffs, N.J.: Prentice-Hall.

Peña, D. and Yohai, V. J. (1995), "The Detection of Influential Subsets in Linear Regression Using an Influence Matrix," *Journal of the Royal Statistical Society (B)*, 57, 145-156.

Rao, C. R. and Mitra, S. K. (1971), *Generalized Inverse of Matrices with Applications*, New York: John Wiley & Sons.

Rencher, A. C. (1995), *Methods of Multivariate Analysis*, New York: John Wiley & Sons.

Romesburg, H. C. (1984), *Cluster Analysis for Researchers*, Belmont, Calif.: Lifetime Learning Publications.

Rousseeuw, P. J. and Leroy, A. M. (1987), *Robust Regression and Outlier Detection,* New York: John Wiley & Sons.

Rousseeuw, P. J. and van Zomeren, B. C. (1990), "Unmasking Multivariate Outliers and Leverage Points (with discussion)," *Journal of the American Statistical Association*, 85, 633-651.

Searle, S. R. (1982), *Matrix Algebra Useful for Statistics*, New York: John Wiley & Sons.

Seber, G. A. F. (1977), *Linear Regression Analysis*, New York: John Wiley & Sons.

Sen, A. and Srivastava, M. (1990), *Regression Analysis: Theory, Methods, and Applications*, New York: Springer-Verlag.

Stewart, G. W. (1973), *Introduction to Matrix Computation*, New York: Academic Press.

Strang, G. (1988), *Linear Algebra and Its Applications*, 3rd ed., San Diego: Harcourt Brace Jovanovich.

Styan, G. P. H. and Puntanen, S. (1994), "A Guide to Books on Matrices and Inequalities, with Statistical and Other Applications," Department of Mathematical Sciences, University of Tampere, Report A 283.

Weisberg, S. (1980), *Applied Linear Regression*, 2nd ed., New York: John Wiley & Sons.

Index

Q

R

S